Life Science

FIFTH EDITION

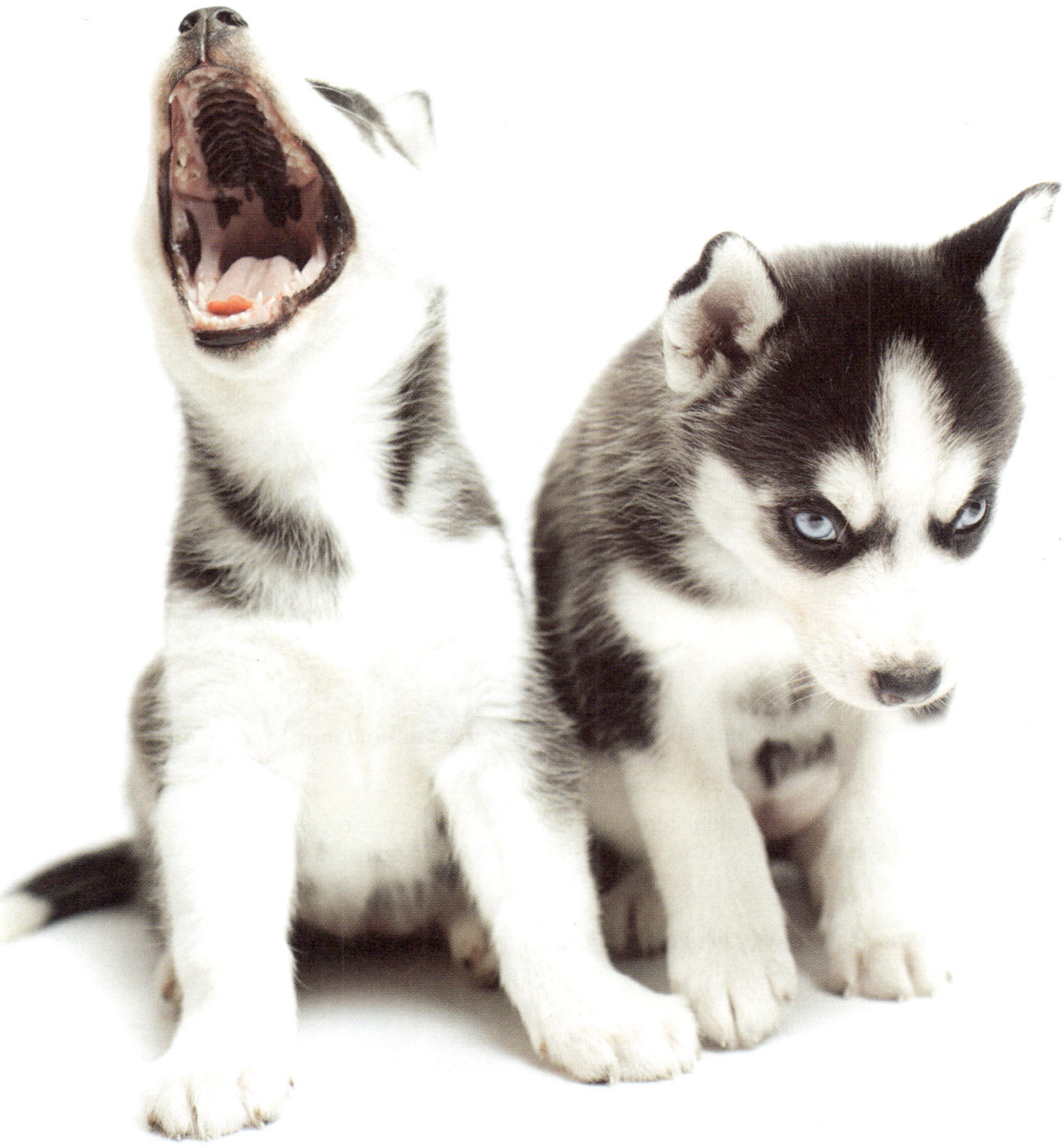

bju press
Greenville, South Carolina

The writers and the publisher have made every effort to ensure that the laboratory activities in this publication are safe when conducted according to the instructions provided. We assume no responsibility for any injury or damage caused or sustained while performing activities in this book. Conventional and homeschool teachers, parents, and guardians should closely supervise students who perform the exercises in this manual.

NOTE: The fact that materials produced by other publishers may be referred to in this volume does not constitute an endorsement of the content or theological position of materials produced by such publishers. Any references and ancillary materials are listed as an aid to the student or the teacher and in an attempt to maintain the accepted academic standards of the publishing industry.

LIFE SCIENCE Student Activities Answer Key Fifth Edition

Coordinating Writer
Rachel Santopietro, MEd

Writers
Christopher D. Coyle
Elwood Groves, MA

Contributing Writer
Jeff S. Foster, MS

Biblical Worldview
Tyler Trometer, MDiv

Academic Oversight
Jeff Heath, EdD
Rachel Santopietro, MEd

Project Editor
Rick Vasso, MDiv

Cover, Design, and Interior Concept Design
Sarah Lompe

Page Layout
Carrie Walker

Illustrator
Sarah Lompe

Permissions
Carrie Hanna
Lily Kielmeyer
Hannah Labadorf
Rita Mitchell
Ashleigh Schieber

Project Coordinators
Chris Daniels
Anthony Every

Formerly published as *LIFE SCIENCE* Lab Manual Teacher's Edition 4th Edition by Jeff S. Foster and Elizabeth A. Lacy.

Photo credits appear on pages 269–70

Answer Key photo credits: **xiv** SpeedKingz/Shutterstock.com; **xvi** AGorohov/Shutterstock.com; **xvii** Photodiem/Shutterstock.com

The front cover photo shows a husky (*Canis lupus familiaris*).

© 2019 BJU Press
Greenville, South Carolina 29609
Fourth Edition © 2013 BJU Press
First Edition © 1984 BJU Press

Printed in the United States of America
All rights reserved

ISBN 978-1-62856-403-7

15 14 13 12 11 10 9 8 7 6 5 4 3 2 1

Contents

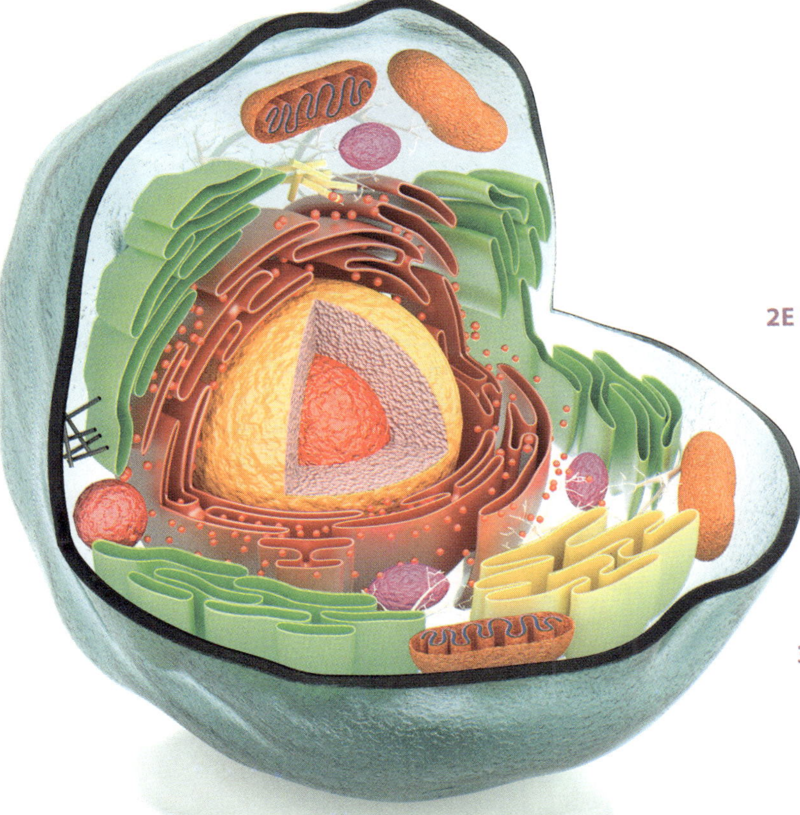

Contents

Contents

Exploring Life

Science—it's a way of learning about God's creation. It's exploring and marveling at the beautiful intricacies of living organisms. And here's the secret: you can learn only so much of that amazing beauty in a textbook. Life science is about doing things! The activities in *LIFE SCIENCE* Student Activities will allow you to explore life science in order to see it more clearly. You'll complete activities that help you learn the concepts you're studying in class, develop laboratory skills, learn to record and interpret experimental data, build problem-solving skills, and much more. And hopefully you'll have fun while you're at it!

LEARNING FROM *LIFE SCIENCE* STUDENT ACTIVITIES

As you look through your Student Activities, you will see two different types of activities—reviews and labs. Each type ties in with your textbook but in different ways.

REVIEWS

Reviews are short pencil-and-paper exercises designed to reinforce the information that you've learned in the textbook. Think of them as extra section reviews to help you master the material and get ready for the test.

LABS

Labs are designed to allow you to investigate some area of life science. While each lab activity is built on information in your textbook, the information in the activity goes beyond the textbook as well. You won't find the right answer in the textbook. In fact, sometimes there is no right answer! The answers to each lab activity will be based on the data that you collect and analyze. Sometimes you will work alone. Other times you will work with a lab partner, in a small group, or with the entire class. You'll learn to use scientific instruments, collect data, and make models. You'll make calculations, access online resources, and perform chemical tests. You will ask questions and then go on to answer those questions. These are things that real scientists do every day.

COMPLETING AN ACTIVITY

When it's time to complete an activity, don't treat it like something to finish as fast as you can. Treat each one as an opportunity to stretch yourself, like a good mental workout.

CONNECT

Connect with what you have learned in your textbook. But even more importantly, connect with the real world. Read the essential and key questions for some real-world context for the lab activity.

SOLVE

As you work through the procedures and questions of each lab activity, you will be using the scientific process and appropriate scientific tools. Your goal is to find the answer to a question or the solution to a problem laid out in the introduction.

EXTEND

Most lab activities have a "Going Further" section. This is an opportunity for you to extend what you've learned in the lab activity to new areas. These activities aren't there just to occupy your time; they're based on solutions to real world issues.

WORLDVIEW IN THE LABORATORY

When professional scientists work in the laboratory, their worldviews always affect what they're doing. When they draw conclusions, their worldviews are hard at work. Their very reason for doing science comes from their worldviews.

As a student scientist, keep your worldview in mind too. When you do an experiment, think about how it relates to a biblical worldview. Some questions will specifically ask you to apply your worldview to the problem. You will be challenged to consciously think like a Christian as you're doing science.

STAYING SAFE

Scientists need to be very serious about safety. No scientist wants to lose his eyesight, experience painful injuries, or be killed while doing an experiment. And no scientist wants to be responsible for injuring someone else when carrying out an experiment.

While the experiments that you'll be doing are designed to be safe, any experiment can injure you if you're not careful. Safety rules help to prevent accidents. But even more important than memorizing safety rules is learning to *think safe*. Think ahead and spot unsafe situations. In other words, *prevent an injury before it happens*. Let's explore some safety strategies.

SPOTTING HAZARDS

When you're in the laboratory, *keep your eyes open*! Look around for things that could cause injuries. These include any objects that could trip someone, chemicals that could splatter, and heat sources that could burn. Before using a sharp tool, think about what else it could cut besides what you're trying to cut. Are you protecting yourself, others, and the cutting surface?

You should also *keep your mind engaged*. Think about how something could break. Ask yourself, "What will happen if I do this or that? Will it fly off in a particular direction? Or will it suddenly become sharp? Is anyone standing in the line of fire?" Remember, you can't be thinking safe if you're messing around, talking, or daydreaming.

POISONS

Poisons aren't found just in murder mysteries. Many chemicals, including those under your kitchen sink, are poisonous. They may not kill you, but they can make you really sick. You won't be using anything very poisonous in these lab activities, but some chemicals could make you ill if you're not careful. How can you protect yourself? Keep chemicals out of your body! Never taste a chemical. Eating or drinking while working with chemicals is obviously a really bad idea. And eating or drinking after working with chemicals can be risky. Always wash your hands after the lab period is over so that you don't unintentionally eat a snack spiced with chemicals. Some chemicals can get into your body through the skin, so wash chemicals off your skin immediately.

BURNS

If you've ever been burned, you know that burns are not fun. And most of the time they're easily preventable. Be alert for anything that gets hot. Burners, stoves, and open flames can burn you very easily. Even more dangerous is hot glass. It may look safe but be *very* hot. Whenever you're doing a lab activity that involves heat, keep asking yourself the question, "Is this hot?" before you touch anything.

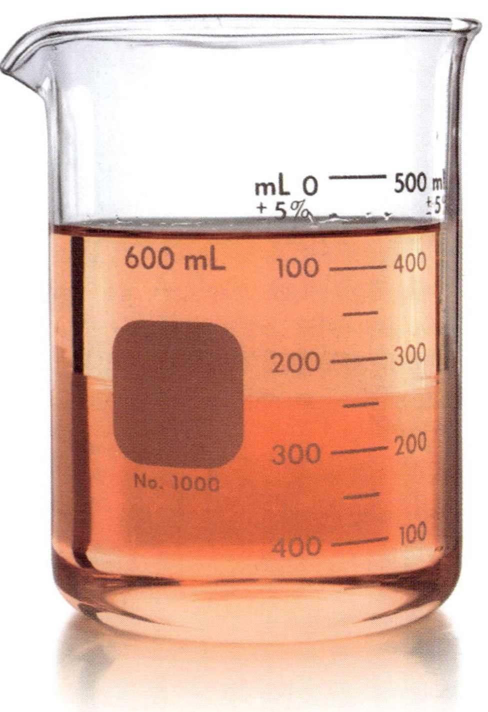

Is this beaker hot? You can't tell just by looking.

Chemicals can burn you too. Strong acids and bases attack your skin and eyes, causing pain and damage. If you ever come in contact with a strong acid or base, wash it off immediately with lots of running water. Laboratories usually have special stations to wash your eyes if chemicals get into them. Know where the eyewash station is in your laboratory.

FIRE AND HEAT

When you're using heat, fire is always possible. Keep flammable items away from heat sources. Know where your laboratory fire extinguishers are and how to use them. Always follow your teacher's rules for fire prevention and evacuation.

Heat things carefully. Containers can crack if they're heated when almost empty. Keep open containers pointed away from you so that you don't get a blast of hot steam in your face. Remember, microwave ovens can be dangerous too. They can make something just as hot as a stove or burner.

FIELD WORK

Going outside may seem safer than being in the laboratory, but there are hazards even in the field. Recognize dangerous places like cliffs and deep water. Watch for areas that might hide venomous animals like snakes or hornets. Learn to recognize some of the most common poisonous plants in your area, such as poison ivy. And be aware of dangerous large animals, whether domesticated, like dogs and bulls, or wild, such as deer or bear. Your eyes are your best defense. Look before you leap, walk, or touch.

Definitely not something you want to bump into!

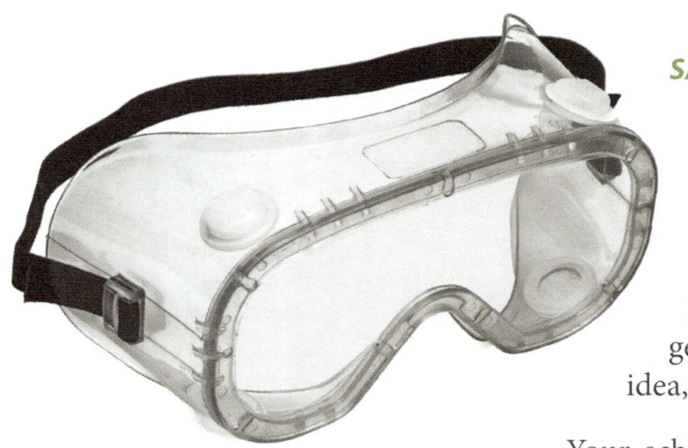

Your lab safety goggles are one of the most important pieces of lab safety equipment.

SAFETY GEAR

Safety gear may seem unnecessary, but it serves an important purpose. So when you're told to wear goggles, protective clothing, or gloves, do it! Goggles may be uncomfortable, but they keep a lot of things out of your eyes. Having two working eyes for a lifetime is definitely worth a few minutes of discomfort! Your lab activity instructions will warn you if protective gear is needed. But anytime you think protective gear is a good idea, use it.

Your school laboratory will probably have other safety features too. These include first-aid kits, eyewash stations, fire extinguishers, and chemical safety data sheets. But your laboratory will definitely come with one very important safety feature—your teacher. When in doubt, ask! If you are uncertain about how to do something, ask. If something spills, tell your teacher. If you get cut or burned, ask your teacher what to do. And if you don't know whether you need to use safety gear, ask first.

SAFETY ICONS

The directions in the investigations include safety icons to alert you to possible danger. The safety icons are explained below. When you see a particular icon in an activity, your teacher will explain to you what the danger is and how to protect yourself.

ANIMALS
Animals that you are to observe or collect may inflict dangerous stings or bites.

BODY PROTECTION
Chemicals, stains, or other materials could damage your skin or clothing. You should wear a laboratory apron, chemical-resistant gloves, or both.

ELECTRICITY
An electrical device (hot plate, lamp, or microscope) will be used. Use the device with care.

EXTREME TEMPERATURE
Extremely hot or cold temperatures may cause skin damage. Use proper tools to handle laboratory equipment.

EYE PROTECTION
There is a possible danger to the eyes from chemicals or other materials. Wear safety goggles.

FIRE
A heat source or open flame is to be used. Be careful to avoid skin burns and the ignition of combustible materials.

PATHOGENS
Organisms encountered in the investigation could cause human disease.

PLANTS
Plants that you are to observe or collect may have sharp thorns or spines or may cause contact dermatitis (inflammation of the skin).

POISON
A substance in the investigation could be poisonous if ingested.

SHARP OBJECTS
Cuts are possible from broken glassware (e.g., broken test tubes, thermometers, or microscope slides) or sharp instruments (scalpels, razorblades, or knives).

You can have a lot of fun participating in life science activities this year. You'll learn fascinating things as you complete the activities. After all, you're studying God's living creation, and there are lots of fascinating organisms in the world! We hope these activities will help increase your knowledge of science as well as your love for science and for the God who is the Creator of all. So get ready to explore the world of life science!

The laboratory experience lies at the heart of science. By its very nature, science is largely empirical, so laboratory skills are essential. It is our goal to develop student scientists, and we hope that this is or will become your goal as well. By building their laboratory skills, students not only experience the gratification of seeing science come alive but also prepare themselves for future science courses.

FOUNDATIONAL CONCEPTS

As your students work through the various activities in this manual, they will encounter three key concepts that illustrate the empirical nature of science. From the very beginning, we emphasize the importance and nature of *data*. Without data, we would not have science. Think of data as the raw material of science. But data is not just collections of numbers. Data can be *qualitative* (descriptive) or *quantitative* (numerical). It can be collected and manipulated in a variety of ways. And data can be good or bad (i.e., valid or invalid), sufficient or insufficient. By making students think about and assess data, we help them develop one of the most important critical-thinking skills needed to navigate the world of modern science. They will see that the way that you interpret data depends greatly on your worldview.

Another crucial concept stressed in this manual is *modeling*. Much of science is about making models. Models are not just structures representing very large or very small objects and systems. They are often numerical, based on data, and represented by visual aids such as graphs. Models describe physical or conceptual phenomena. Significantly, they offer *predictive power*. Indeed, models are evaluated and judged on the success of their predictions. The scientific community relies heavily on models and bases many of its conclusions on them. As your students create their own models from real data, they will understand where scientific predictions and conclusions come from. They will also explore the serious topic of *model limitation*. Even professional scientists can fail when they draw unjustified conclusions from limited models.

Finally, your students will develop the crucial manual skills without which reliable science could not happen. They will become adept at measuring events and properties, recording and graphing data, and correctly using scientific tools. What they learn in this class will pay dividends when they move on to high-school and college science courses.

As students perform these labs under your guidance, keep these three key concepts in mind at all times. Students often rush labs, seeing them as something to get through as quickly as possible. Help them resist that tendency. Encourage them to think about the importance of their data and their models. Help them to become skillful and accurate as they use new and unfamiliar tools. Above all, show them the joy that comes from science as they discover new intellectual vistas that reveal the wonders of the Creator!

SPECIAL FEATURES

The *LIFE SCIENCE* Student Activities 5th Edition (SA) has been updated and represents a significant enhancement in teaching life science labs. About a quarter of the lab activities are new to this edition, and those retained from earlier editions have been revised or enhanced.

Each chapter includes reviews that connect to particular sections of the textbook. These are pencil-and-paper activities that reinforce the information the student has learned in that section. They also provide opportunity for differentiated learning, since they can be used as extra practice or enrichment, depending on the need. The objectives in the reviews repeat objectives in the corresponding section in the Teacher Edition.

One new feature in this edition is the addition of inquiry labs. These lab activities are not your standard step-by-step type labs. Rather they outline a particular issue or problem and then allow the students to investigate as they desire. The lab instructions in the SA are limited by design, but we give plenty of support and guidance in this *LIFE SCIENCE* Student Activities Answer Key (SA-AK). You will notice on some of these inquiry labs that there are additional pages added to the SA-AK to provide you the background and support you need to guide students through the activities.

We have also added one STEM (Science, Technology, Engineering, and Mathematics) lab activity in Chapter 12. This activity will guide students through the design process as they design, construct, test, and modify an insect trap. As with the inquiry labs, we provide teachers with the support they need in expanded pages in the SA-AK.

Since graphing offers a simple but powerful way to model, students will construct graphs in many lab activities. But graphing often frustrates students as they struggle to choose suitable axis scales on an intimidatingly blank sheet of graph paper. All of us probably remember finishing up with a tiny graph crammed into one corner of the sheet! However, commonly available software makes setting up the graph much easier, leaving the student more time to analyze and interpret the graph. The student should have access to a spreadsheet program such as Microsoft Excel®. This will help students create their models quickly and efficiently and then move on to draw thoughtful conclusions from them.

Lastly, with mainstream science becoming so dominated by naturalistic worldviews, we believe that it is crucial to cultivate a strong biblical worldview. Many of the lab activities stress the points at which worldviews come into conflict. For example, the very last lab activity explores human population growth trends from the twentieth century. Many of those holding to a naturalistic worldview feel that this situation is dangerous and that the world should have far fewer people. We contrast this position with the Christian worldview that regards population growth as desirable and biblical (Gen. 1:28) and sees science as a tool to make it possible. We also lead the students to recognize that predictive models become less accurate as the time between the data pool and the prediction increases.

USING THE TEACHER NOTES

In this SA-AK, each lab activity includes numerous teacher notes in the margins. These provide additional guidance, clarify equipment requirements, or demonstrate calculations and other processes. Read them over before you perform the lab for a clearer understanding of the process. Many of your questions are anticipated and answered by the notes.

Some notes provide extra guidance for worldview-critical situations in which you will be dealing with topics that are especially philosophical. Be sure to read these over thoroughly so that you will be ready to handle these crucial lab activities. Most notes are flagged by special icons to emphasize their purpose. The key below explains the meaning of each icon.

BEST PRACTICES

This icon indicates notes that include suggested ways to apply research-based educational strategies in your classroom. These strategies include group activities and discussions, activating prior knowledge, active learning, and formative assessments.

BIBLICAL WORLDVIEW SHAPING

Notes flagged with this icon introduce discussions or ideas that reinforce a Christian worldview of Life Science.

DEMONSTRATION

This icon will alert you to activities that may work well as a teacher demonstration or to demonstrations that might aid students as they perform the activity.

DIFFERENTIATED INSTRUCTION

This icon alerts you to procedures that may be beneficial for some students. These may include lab procedures that may be difficult for students with a disability. The note will include suggestions to alleviate the problem. Other notes may suggest methods of instruction to help students with various learning styles.

DIFFICULT CONCEPT

This icon alerts you to concepts that students may find difficult to grasp or that may be a common misconception. Information and suggestions will be given on how to clarify or reinforce such material.

EQUIPMENT

Notes flagged with this icon provide extra information about the equipment and materials needed for the lab activity. Acceptable substitutions, local sources, or preparation instructions are included.

HELPFUL TIP

Helpful tips provide you with hints and ideas for coordinating and managing the lab activity so that students get better results or have a smoother experience.

MATH

The math icon signals a note that provides guidance for calculations or other math-related tasks. Worked-out problems are often flagged by this icon.

RESOURCES

Notes with this icon direct you to a resource that will help you in completing the lab activity. The note often directs you to a link on TeacherToolsOnline.com or suggests that you search for information using your favorite search engine and the keywords provided.

SAFETY

When you see this icon, be sure to read the note since it contains crucial information that will help you to be safe or to avoid a potential safety problem.

To the Teacher

SCHEDULING LAB ACTIVITIES

The SA contains more activities than you can perform in an average school year. We purposely designed it that way to give you choice and flexibility. We also tried to create a variety of activities suitable for different educational settings.

Try to do at least one lab activity for each chapter you cover. While some labs can be performed with very basic equipment, others require well-equipped laboratories. This allows you to adapt the SA to the needs of your students and the resources of your school. Since there is something for everyone, you should have no trouble selecting suitable labs. LogosScience is a company that has built science kits to accompany the SA. You can order these through the BJU Press website.

The reviews can be assigned in class or as homework. Use them as needed and in the manner that best fits the needs of the students in your class.

We typically offer two lab activities for each chapter, though a few chapters have only one. Choose them carefully and adapt them to your particular teaching situation. For lab activity scheduling suggestions, see the *Life Science* Teacher Edition 5th Edition Lesson Plan Overview.

PERFORMING LAB ACTIVITIES

The lab activities in the SA follow a logical linear flow. Each one opens with a brief, thought-provoking passage that introduces the topic and sets the stage for the activity. Students should always read the introduction carefully. You may also want to have a brief class discussion before beginning certain labs, particularly those that deal with complex or controversial topics.

Questions (these are numbered) are intermingled with lab procedure steps (these are numbered *bullets*). By answering questions as they go along, rather than all at once at the end, students will think more deeply about what they have just done. Answers to some questions logically suggest the need for further exploration and introduce the next procedure step.

Encourage students to answer questions thoroughly and completely. Most yes/no answers require an explanation. When you grade lab activities, insist on clear, coherent answers that address everything requested in the question. Expect accurate numbers, clear thinking, and complete sentences where appropriate. Following directions and expressing one's ideas clearly are crucial skills for all phases of life.

As a general rule, the students will find data tables or diagram space at the end of each lab. This allows students to pull those pages out to work on while still being able to read the lab procedures and questions. In cases where it helps students, tables, graph paper, and diagram space may be located within the lab procedures and questions.

Carefully supervise steps involving lab equipment, both to ensure safety and to help cultivate the new skills that students are developing. Monitor calculations and modeling steps. Students will come to see these steps as interesting and gratifying rather than as drudgery if they understand them and do them well.

Consider conducting class discussions after certain lab activities, particularly those that address worldview issues. These lab activities stress ideas that are vital to a developing Christian and are well worth the extra time investment.

EQUIPMENT AND MATERIALS

Each lab includes a list that specifies the equipment and consumable materials required to perform the lab activity. Most of these items are found in basic school laboratories. Many items can be obtained locally. Very few are exotic or expensive. In the inquiry labs, we intentionally tend to list little, if any, equipment in the SA to avoid influencing the direction that students take in the lab activity. For those activities, we have provided a suggested list of equipment in this SA-AK.

To make planning your upcoming school year easier, we have included both *alphabetical* and *by-lab* equipment lists at the end of this book. These lists include the equipment needed, the size needed, and quantities required for each student workgroup. In the alphabetical list, you can find the lab in which each equipment item is used. Where appropriate, we include suggestions for substitutions and local sources. As you work out your lab schedule for the year, use these lists to help you decide what to order. Many schools like to order all equipment and supplies at a single time. By being well prepared, you will avoid having an equipment crisis mid-year!

SAFETY

Laboratory safety prevents an enjoyable school year from being punctuated by tragedy. No teacher wants to have students injured or property damaged, particularly if a situation is preventable. Following good lab safety practices can significantly reduce the likelihood of such an outcome.

The SA takes a slightly different approach to safety than you may have encountered before. Instead of having students memorize a list of safety rules, we suggest that you train them to *think safe*. The key idea behind this method is to cultivate a mindset of always looking for unsafe situations and preventing them from developing. To their advantage, students will start thinking ahead and paying closer attention to what they are doing and so will be better prepared when something unexpected occurs.

To help you develop this safety mindset in your students, the introduction to the SA includes a section that stimulates thinking along these lines. We strongly suggest that you devote your first lab period to a safety discussion. Instead of just listing rules, spend time brainstorming with your students about unsafe situations and ways to prevent them from happening. Some of the scientific supply companies have great safety demonstrations that may be useful in this first lab activity. Obviously, your laboratory needs rules; more importantly, it needs alert and thinking students.

You, of course, are the most important safety feature in your laboratory, more crucial than the fire extinguisher, eyewash station, or emergency shower. It is you who provide the maturity, expertise, and calmness needed to avoid accidents and to deal with them when they do occur. If you are a new teacher or have limited science-teaching experience, spend some time reading about school laboratory safety practices. There are many excellent online resources published by reputable safety organizations. You may also benefit from seeking the help of an experienced science teacher. Remember, if you are unsure of how to perform a task safely, wait until you *are* sure before doing it.

The lab activities in the SA involve a limited number of chemicals, most of them comparatively safe. But any chemical can be a hazard under the right conditions. Be sure that you understand the proper storage, handling, and disposal procedures for each substance in your laboratory.

Maintain an up-to-date file of Safety Data Sheets (SDS). If you order a chemical and it does not come with one, contact the manufacturer or supplier. Many suppliers offer free downloadable SDS files on their websites. Read over the SDS for each new chemical that you acquire.

Emphasize to your students that we never eat or drink in the laboratory. As with any safety practice, you can do the greatest benefit—or harm—by your example. Set the good example by never eating or drinking in the laboratory.

Confirm that your laboratory has adequate eye, hand, and clothing protection. Never work through a lab activity that requires protective gear unless you have enough for everyone. Be sure that your students know where the emergency equipment (fire extinguishers, eyewash stations, fire blankets, first aid kit, etc.) is located.

LET THE ADVENTURE BEGIN!

As you prepare for your year of teaching life science, step away from the mechanical and logistical details long enough to remember that science is a fascinating and enjoyable field of study. This attitude is infectious! If you can communicate it to your students, they will absorb it and respond in kind. Teaching science is not about schedules, assignments, and well-maintained labs. It is about connecting with students and helping them to think in new and unfamiliar ways. You are the one who will make science a bright and exciting adventure—or a tedious and plodding ordeal. Take advantage of your opportunity to show them the wonders of their Creator!

REVIEW 1A

DISTINGUISHING DATA FROM INTERPRETATION

NAME:

DATE:

Read each of the following statements. In the space provided, mark whether each statement is evidence (E), a biblical interpretation of evidence (B), or a naturalistic interpretation of evidence (N).

E **1.** No dinosaurs appear to be alive today.

N **2.** Dinosaurs became extinct about 65 million years ago.

E **3.** Scientists find dinosaur fossils all over the world.

B **4.** Dinosaur fossils were laid down during a worldwide flood.

E **5.** The skeletons of modern birds are similar to dinosaur skeletons in some ways.

N **6.** Because bird skeletons are similar to dinosaur skeletons, we can say that modern birds evolved from dinosaurs millions of years ago.

N **7.** Humans are related to apes and other primates.

B **8.** Humans are distinct from other organisms because they are God's image bearers.

B **9.** Suffering and death exist because of mankind's sin.

B **10.** People have an obligation to help and serve others.

E **11.** Earth is the only planet where life has been found.

N **12.** Ancient humans were more primitive than their modern counterparts.

N **13.** All life will cease when the universe eventually runs out of energy.

B **14.** We should do science to help people flourish.

REVIEW 1A OBJECTIVE

- Compare a naturalistic worldview with a biblical worldview. **BWS**

REVIEW OBJECTIVES 

The objectives for the review exercises are applicable section-level objectives from the *Life Science* Teacher Edition 5th Edition. They are included here to help focus the review.

DISTINGUISHING DATA FROM INTERPRETATION 1

NOTES

DESCRIBING LIFE

NAME:

DATE:

Directions

Complete the concept definition map about what life is by filling in the four boxes with words or phrases that answer the questions about life.

WHAT IS IT?

an organism with all of a set of certain characteristics that responds to internal and external changes to maintain homeostasis

WHAT ARE ITS CHARACTERISTICS?

made of cells, needs energy, grows, responds to its environment, can reproduce

LIFE

WHAT ARE SOME EXAMPLES?

Answers will vary.

Examples: guinea pig, lizard, goldfish, parakeet, frog, grasshopper

WHAT IS IT NOT?

Answers will vary but may include any inanimate object or idea; it may also include dead organisms.

Examples: statue, computer, car, old log, rock

REVIEW 1B OBJECTIVE

- Give evidence that something is alive using the characteristics of life.

CONCEPT DEFINITION MAP EXAMPLE 

You may need to refer students to page 445 of Appendix B in *LIFE SCIENCE* Student Edition 5th Edition to show them what a completed concept definition map looks like.

NOTES

FORMING A HYPOTHESIS

NAME:

DATE:

Before a scientist can perform an experiment, he must define the scientific problem and form a hypothesis. The problem is usually expressed in the form of a question. Its purpose is to limit the scope of the experiment. A hypothesis is a proposed answer to the problem. It describes the relationship between the variables being studied and is usually based on some research. During the experiment, the independent variable is changed to observe its effect on the dependent variable. The hypothesis can be stated in an if-then format. For example, "If the bean plants are given extra fertilizer, then they will produce more beans." In this case, differing amounts of fertilizer are watched for their effect on the number of beans grown. Both variables are easy to observe. Try your hand at writing hypotheses like a life scientist does.

Directions

For each of the following topics and problem statements, give the possible variables and write a testable hypothesis relating the two variables.

EXAMPLE TOPIC: PLANT GROWTH

Problem: Is the growth of plants related to the amount of fertilizer they receive?

1. Independent variable: *amount of fertilizer*

2. Dependent variable: *height of plants*

3. Hypothesis: *If the growth of plants is related to the amount of fertilizer they are given, then plants that receive more fertilizer will grow taller than plants that receive less fertilizer.*

TOPIC: PLANT GROWTH

Problem: Is the color of light related to plant growth?

1. Independent variable:

   the color of light

2. Dependent variable:

   the height of plants

3. Hypothesis:

   Answers will vary. Example: Orange light will cause the plants to

   grow the tallest.

REVIEW 1C OBJECTIVE

- Describe the scientific process.

GIVING YOUR STUDENTS CHOICES

You may consider allowing students to choose which scientific topics to explore on the basis of their interests. Consider having them do three topics from this first section and two from the Ramping It Up section on page 7, where they will also write the scientific problem.

HYPOTHESES

While a hypothesis does not have to be written in if-then format to be correct, it is good for students to do so as they learn to write valid, testable hypotheses.

STUDENT HYPOTHESES

The hypotheses listed in this first section are examples. Student answers will vary. Grade the hypotheses on the basis of their testability and relationship to the variables. The hypotheses given are the "then" portion of the if-then statements.

NOTES

Problem: Does temperature affect how quickly sugar dissolves in water?

4. Independent variable:

   the temperature

5. Dependent variable:

   the speed at which sugar dissolves

6. Hypothesis:

   Answers will vary. Example: As the temperature of the water

   increases, the sugar will dissolve faster.

TOPIC: BACTERIAL GROWTH

Problem: Does temperature affect the growth of bacteria?

7. Independent variable:

   the temperature

8. Dependent variable:

   the growth of bacteria

9. Hypothesis:

   Answers will vary. Example: Bacteria will grow best in a particular

   temperature range.

TOPIC: TEST SCORES

Problem: Are test scores related to the frequency of studying?

10. Independent variable:

   the frequency of studying

11. Dependent variable:

   test scores

12. Hypothesis:

   Answers will vary. Example: More frequent studying will result in

   higher test scores.

TOPIC: POPCORN POPPING

Problem: Is the price of popcorn related to the percentage of kernels that pop?

13. Independent variable:

   the price of popcorn

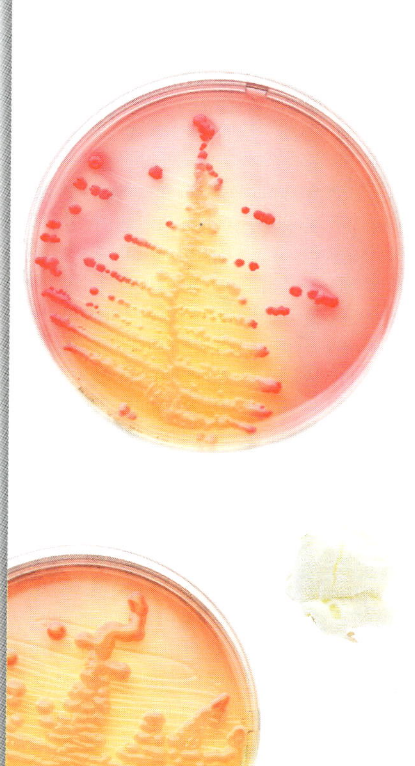

14. Dependent variable:

   the percentage of kernels popped

15. Hypothesis:

   Answers will vary. Example: Higher-priced popcorn will have a

   higher percentage of kernels popped.

TOPIC: EXERCISE AND MEMORY RETENTION
Problem: Is there a relationship between memory retention and exercise?

16. Independent variable:

   the frequency of exercise

17. Dependent variable:

   memory retention

18. Hypothesis:

   Answers will vary. Example: Exercising more often will increase

   memory retention.

Ramping It Up
Now try your hand at writing a problem! What would you investigate if you were interested in experimenting on the following topics?

TOPIC: SMELL AND TASTE

19. Problem:

   Answers will vary. Example: Is the intensity of the taste of food

   related to how strong it smells?

20. Independent variable:

   the strength of the smell of the food

21. Dependent variable:

   the intensity of taste

22. Hypothesis:

   Answers will vary. Example: The taste will be more intense for

   foods that have a stronger smell.

TOPIC: CRICKET CHIRPING

23. Problem:

   Answers will vary. Example: Is there a relationship between the

   rate of cricket chirping and the air temperature?

24. Independent variable:

air temperature

25. Dependent variable:

the rate of chirping

26. Hypothesis:

Answers will vary. Example: The chirping rate of crickets will increase as the temperature increases.

TOPIC: ATTRACTING HUMMINGBIRDS

27. Problem:

Answers will vary. Example: Do certain colors of hummingbird feeders attract hummingbirds more than other colors?

28. Independent variable:

the color of the hummingbird feeder

29. Dependent variable:

the number of birds that visit the feeder

30. Hypothesis:

Answers will vary. Example: Red will attract more hummingbirds than yellow, green, or blue will.

TOPIC: ICY ROADS

31. Problem:

Answers will vary. Example: Which treatment melts the ice on icy roads the fastest?

32. Independent variable:

the type of road treatment

33. Dependent variable:

how fast the ice melts

34. Hypothesis:

Answers will vary. Example: A brine (salt) solution will melt ice faster than sand or charcoal cinders will.

LAB 1D

LOOKING WITHIN

Observing Cells with a Microscope

HOW CAN I SEE CELLS?

In this chapter we've investigated the properties of life, including cells. We've studied how microscopes work. But how can we see cells? You can use a microscope to find out what cells look like.

Procedure

You will use a microscope to observe several prepared slides (slides that have preserved tissues mounted on them) and fresh specimens that your teacher has chosen. Prepared slides usually have specimens that have been stained for easier viewing. Different organisms have different kinds of cells. Be sure to observe at least one kind of animal cell and one kind of plant cell. View your slides at both low and high magnification.

USING THE MICROSCOPE

Always follow these instructions to properly obtain, set up, and return your microscope.

Always carry the microscope properly.

Excessive jarring and bumping may bring the lenses out of adjustment. To avoid damaging the microscope, observe the following rules.

- If it is necessary to take a microscope out of a cabinet or cupboard, be careful not to bang the microscope against the sides of the cabinet.

- Carry the microscope with one hand under the *base* and the other hand holding the *arm* of the microscope.

- Be sure to keep the microscope close to your body in an upright position so that the *eyepiece* does not slip out of the *eyepiece tube* or *body tube* (depending on the style of microscope).

- Gently place the microscope on the table and position it about 8 cm from the edge.

NAME:

DATE:

Key Questions 

- What do cells look like?

- How do I use a microscope?

Equipment

microscope

prepared slides of various kinds of cells

slides of fresh specimens

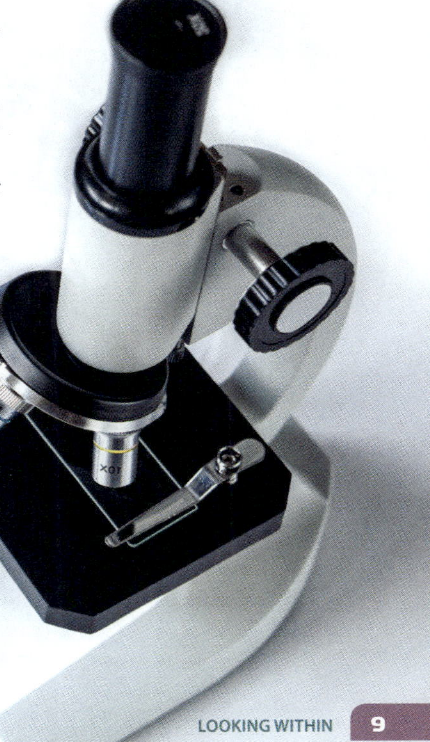

LAB 1D OBJECTIVES

- Correctly use a light microscope to view an image.

- Practice scientific sketching.

- Describe differences between plant and animal cells.

SUGGESTIONS FOR PREPARED SLIDES 

The prepared slides that you provide for your students may be purchased from a supply company or made from fresh specimens before class. Suitable fresh specimens include edges of elodea leaves, geranium leaf epidermis, flower petal epidermis, stained onion skin, live protozoans (e.g., amoeba, euglena), cheek epithelial cells, banana cells, potato cells, and tomato skin.

PREPARING SPECIMENS 

Advanced students can benefit from learning how to prepare their own slides. For instructions on how to do this, see Appendix C in the Student Activities.

LOOKING WITHIN   9

NOTES

Prepare the microscope properly.

You may need to clean your microscope before using it. Observe the following rules while you clean it.

- Use only lens paper to clean lenses and glass surfaces. Wipe the lenses in one direction across the diameter of each lens.

- Consult your teacher if any material remains on your eyepiece or objectives.

- Under no circumstances should you attempt to take your microscope apart!

Return the microscope properly.

When you have finished using the microscope, you should observe the following rules.

- If the microscope has a built-in light source, be sure that the switch is in the off position.

- Make sure that the body tube is straight up and down (if your model is one that has a tilting arm).

- Position the lowest-power objective directly under the body tube.

- Adjust the body tube to its lowest position.

- Carefully return the microscope to the place where you obtained it.

CALCULATING THE MAGNIFYING POWERS ON A MICROSCOPE

A microscope works by passing light (either from a lamp in the base of the microscope or reflected from another source by a mirror) up through a specimen and into the lenses of the microscope. The amount of light that passes through the specimen may be regulated by a *diaphragm* under the stage.

The first set of magnifying lenses that the light passes through is in the *objective*. Most microscopes have several objectives of differing powers (amounts of magnification). The number (10×, 40×, etc.) on each objective tells you how many times larger the image of the specimen will appear after it passes through the objective. The objective you choose can be properly positioned by rotating the *nosepiece* until the objective is pointing down over the specimen.

After passing through the objective lenses, the light passes through the *eyepiece* (ocular) lenses. Most eyepieces have a magnification power of 10×, which makes the image another ten times larger. To find the total magnification of a microscope, multiply the powers of the eyepiece and the objective that you are using. For example, if you use a 40× objective and a 10× eyepiece, the image in your microscope will be magnified 400 times (40× times 10× equals 400×).

❶ Compute the magnifying powers on your microscope.

❷ Fill in Table 1 regarding your microscope. Your microscope may have only two or three objectives. If so, use only the first two or three lines of the table.

Table 1

	Objective power	×	Eyepiece power	=	Total magnification
Objective 1	__________×		__________×		__________×
Objective 2	__________×		__________×		__________×
Objective 3	__________×		__________×		__________×
Objective 4	__________×		__________×		__________×

In this lab activity, the term *low power* refers to a total magnification of about 100×. The term *high power* refers to a total magnification of about 400×.

VIEWING A SPECIMEN ON LOW POWER

Now you'll observe several slides of fresh and preserved specimens that your teacher has prepared. Some fresh specimens must be stained to be seen clearly; others have natural color and do not need to be stained.

Prepare your microscope for viewing.

❶ Use the *coarse focus knob* to raise the nosepiece to its highest setting (or lower the *stage* to its lowest setting).

❷ Open the diaphragm on your microscope to its largest setting.

❸ If your microscope has an electric light source, plug it in and turn it on.

❹ If your microscope has a mirror, adjust it until you are able to see a bright light through the eyepiece (ocular). You should try to use the brightest light source available, but you should not use direct sunlight.

Place a microscope slide for viewing.

❶ Obtain a specimen from your teacher. Handle the slide only by the edges. This will prevent you from viewing fingerprints by mistake.

❷ Place the slide on the stage of your microscope and position it so that the specimen is centered over the opening in the stage. Clip the slide in place.

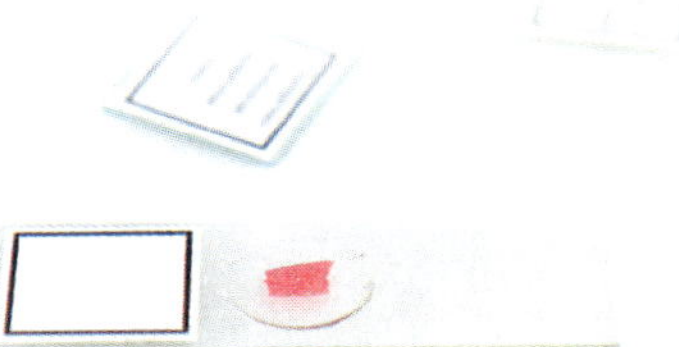

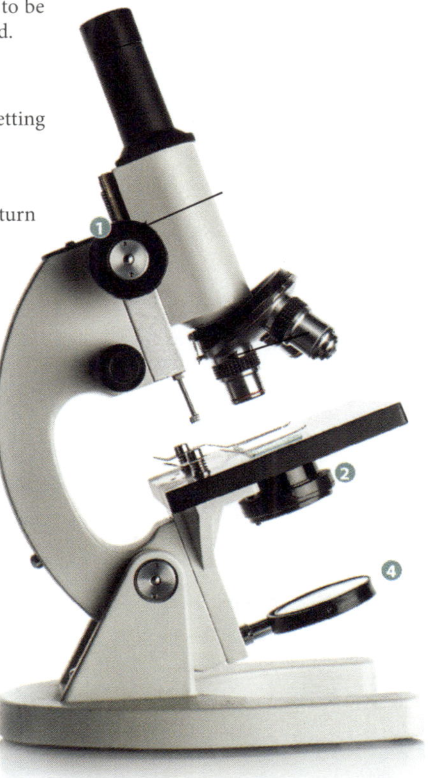

WORKING WITH SLIDES 

Even after students have brought their specimens into focus, they still may not have a good sample for viewing. Be sure to coach your students on how to carefully maneuver a slide while viewing it. Some microscopes have controls for moving the entire stage around, but student scopes usually require slides to be moved manually.

If your students have never used a microscope before, they may be initially perplexed by the movement they see through the eyepiece while moving a slide on the stage. The image they are seeing is inverted; students will need to move the slide in the opposite direction from that which they desire the image to move. For example, if they want the image to move to the right, they need to move the slide to the left. It takes some getting used to.

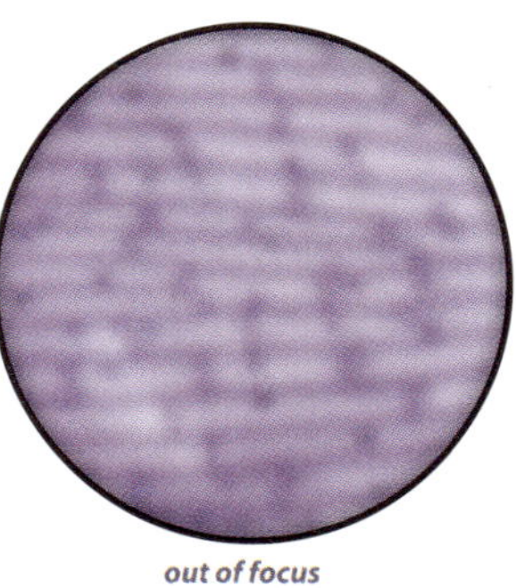

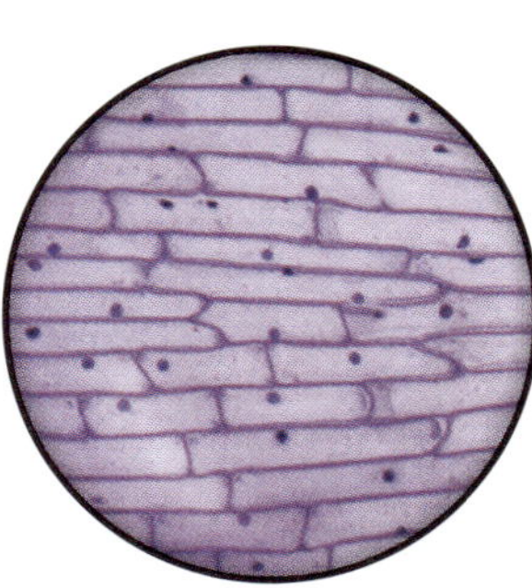

CHANGING TO HIGH POWER 

The instructions here are for use with parfocal microscopes, the typical type of student scope in use today. Parfocal scopes are designed to need a minimum of refocusing when switching from low to high power. You should verify that your scopes are of this type before starting the activity. If your students are using non-parfocal scopes, they will need to raise the body tube (or lower the stage) slightly before changing objectives in order to avoid damaging slides and lenses.

Focus the microscope on low power.

1 Position the low-power objective so that it is directly under the body tube of your microscope. On most microscopes the objectives will click into place.

2 While looking at the side of your microscope, turn the coarse adjustment knob until the objective almost touches the cover slip.

3 Look through the eyepiece and slowly turn the coarse adjustment knob so that the body tube goes up (or the stage goes down) until you are able to see the specimen.

(*Note*: If you have raised the objective more than an inch from the cover slip, you have raised it too far and need to repeat Steps **2** and **3**.)

Never turn the coarse adjustment knob so that the body tube goes down (or the stage goes up) while you are looking through the eyepiece! You could cause the objective to push into the slide and break it.

If you are still unable to see the specimen, try the following.

- Check to make sure that the specimen is directly above the center of the opening in the stage and directly under the objective.

- Adjust the diaphragm to a slightly smaller setting and then try to focus the microscope again.

If you still cannot see the specimen, ask your teacher for help.

4 Once you have found the specimen by using the coarse adjustment knob, use the *fine adjustment knob* to obtain a clear image. You can adjust the fine adjustment knob in either direction, but you should never have to turn it more than a full turn either way.

Note that you often may need to readjust your fine adjustment knob while you are viewing something through the microscope.

VIEWING A SPECIMEN ON HIGH POWER

Some specimens are best viewed under low power. But sometimes it is helpful to view a specimen under higher magnification. Follow these steps to see your specimen under high power.

- Make sure that your specimen is first in focus under low power. Carefully rotate the nosepiece to move the high power objective into place.

- Refocus if necessary using the fine adjustment knob.

MAKING SCIENTIFIC DRAWINGS

One of the best ways to document and understand some of the complex biological structures that you observe in this class is to make a scientific drawing. Drawing is a great field skill to have. As you draw, concentrate on the shape and textures of the different structures that you observe. By the time you have finished, you should better understand the structures that you have drawn.

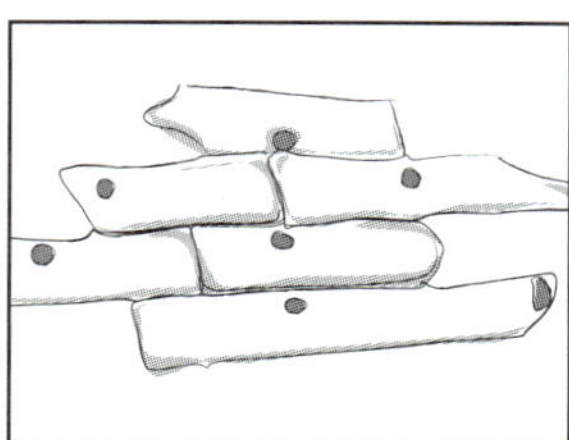

- Center the name of the specimen above the drawing, including the part that you are observing.
- If you are using a fresh specimen, note this on your drawing also.
- Make the drawing large and center it in the space given.
- Use a pencil for most drawings so that you can make changes. You can use colored pencils if doing so helps you remember and understand your drawing better.
- Add labels after the drawing is complete. Label only what you observe.
- For sketches of microscope images, indicate the power in the lower right-hand corner. Don't draw the microscope field.

❶ In the spaces provided, sketch the cells on your specimen slides as you observe them.

❷ Briefly describe the size, shape, color, and number of cells seen.

NAME: ___________________

Type of cells observed:	**Type of cells observed:**
Type of cells observed:	**Type of cells observed:**

Summing Up

1. Describe any differences that you observed between animal and plant cells.

   Answers will vary. Students with prior knowledge of cell structure may note structures such as cell walls, central vacuoles, and chloroplasts that are present in plant cells but not in animal cells.

2. Did you observe any fresh specimen cells moving? If so, describe what you saw.

   Answers will vary. If live protozoans are observed, students may observe flagella, cilia, or amoeboid movement.

3. What other observations were you able to make regarding cells?

   Answers will vary.

LAB 1E

THIS OR THAT?

Using a Key to Classify Organisms

HOW CAN I CLASSIFY UNFAMILIAR ORGANISMS?

When Michael Fay hiked 3000 miles through the tropical rainforests of Gabon in 1999, he stumbled upon many organisms, some of which he didn't know at first glance. How could he identify them, and how could he know whether he had discovered a new organism?

Scientists like Michael Fay use dichotomous keys. *Dichotomous keys* are flow charts used to classify organisms according to physical characteristics that can be observed. One of these characteristics is *symmetry*. An organism shows symmetry if it can be divided into equal halves. If an organism has a top and bottom but no right and left sides, it displays *radial symmetry*. Another characteristic is a skeleton. If an animal doesn't have a hard outer covering, then it must have some other type of support. Often it has a skeleton inside, including a backbone. Look for other features to classify organisms, such as gills, feathers, hair, tentacles, shells, or body segments.

Now you get to do it! Using the Dichotomous Key on page 17, identify the phylum, subphylum, or class to which each animal in the picture belongs. Start at the first row, and follow the directions. Sometimes you'll get a few hints along the way. Assume that animals have backbones unless otherwise stated.

NAME:

DATE:

Key Questions ?

- What is a dichotomous key?
- How can I use a dichotomous key to classify unfamiliar organisms?

Equipment E
Dichotomous Key

Michael Fay

1. has no backbone

class Bivalvia

2.

class Amphibia

3. has no backbone

phylum Porifera

4. has no backbone

phylum Cnidaria

LAB 1E OBJECTIVE

- Define *dichotomous key*.
- Classify organisms using a dichotomous key.

VISUALIZING MICHAEL FAY'S MEGATRANSECT 

For more information about Michael Fay's transect, do a keyword search for "megatransect" or "Michael Fay transect." Look for a short video to show your class to put this lab activity in context.

GIVING A GRADE? 

This activity requires students to rely on prior knowledge or to identify structures not yet covered. Therefore, you may consider *not* giving students a grade that is based on the accuracy of their answers. Alternatively, you can offer this as a bonus grade that is based on accuracy.

WORKING THROUGH THE SOLUTIONS 

1. clam (The sequence students should take in identifying it is 1 → 6 → 7 → 9 → 14 → 17 → class Bivalvia.)
2. salamander (1 → 2 → 3 → 4 → class Amphibia)
3. sponge 1 → 6 → phylum Porifera)
4. jellyfish (1 → 6 → 7 → 8 → phylum Cnidaria)

NOTES

5. tapeworm (1 → 6 → 7 → 9 → 14 → 15 → phylum Platyhelminthes)

6. millipede (1 → 6 → 7 → 9 → 10 → 11 → 12 → class Diplopoda)

7. octopus (1 → 6 → 7 → 9 → 14 → 17 → 18 → class Cephalopoda)

8. bird (1 → 2 → class Aves)

9. tick (1 → 6 → 7 → 9 → 10 → 11 → 13 → class Arachnida)

10. sea urchin (1 → 6 → 7 → 8 → phylum Echinodermata)

11. crayfish (1 → 6 → 7 → 9 → 10 → subphylum Crustacea)

12. snail (1 → 6 → 7 → 9 → 14 → 17 → 18 → class Gastropoda)

13. leech (1 → 6 → 7 → 9 → 14 → 15 → 16 → phylum Annelida)

14. bat (1 → 2 → 3 → class Mammalia)

5. has no backbone

phylum Platyhelminthes

6. has no backbone; has two pairs of jointed legs per body segment

class Diplopoda

7. has no backbone or radial symmetry

class Cephalopoda

8.

class Aves

9. has no backbone

class Arachnida

10. has no backbone or tentacles; has radial symmetry

phylum Echinodermata

11. has no backbone

subphylum Crustacea

12. has no backbone or tentacles

class Gastropoda

13. has no backbone; has a segmented body

phylum Annelida

14. has hair but not feathers

class Mammalia

Dichotomous Key to Identification of Animal Phyla and Classes

1. Does it have a backbone?	**Yes**	Go to number 2.
	No	Go to number 6.
2. Does it have feathers?	**Yes**	class Aves
	No	Go to number 3.
3. Does it have hair?	**Yes**	class Mammalia
	No	Go to number 4.
4. Does it have scales?	**Yes**	Go to number 5.
	No	class Amphibia
5. Does it have gills?	**Yes**	class Osteichthyes
	No	class Reptilia
6. Does it have symmetry?	**Yes**	Go to number 7.
	No	phylum Porifera
7. Does it have radial symmetry?	**Yes**	Go to number 8.
	No	Go to number 9.
8. Does it have tentacles or stinging cells?	**Yes**	phylum Cnidaria
	No	phylum Echinodermata
9. Does it have jointed legs?	**Yes**	Go to number 10.
	No	Go to number 14.
10. Is the first pair of legs modified into large pincers?	**Yes**	subphylum Crustacea
	No	Go to number 11.
11. Does it have more than eight legs?	**Yes**	Go to number 12.
	No	Go to number 13.
12. Does it have two pairs of legs per body segment?	**Yes**	class Diplopoda
	No	class Chilopoda
13. Does it have only six legs?	**Yes**	class Insecta
	No	class Arachnida
14. Does it have the body of a worm?	**Yes**	Go to number 15.
	No	Go to number 17.
15. Does it have a flattened body?	**Yes**	phylum Platyhelminthes
	No	Go to number 16.
16. Does it have a segmented body?	**Yes**	phylum Annelida
	No	phylum Nematoda
17. Does it have a two-part shell?	**Yes**	class Bivalvia
	No	Go to number 18.
18. Does it have tentacles?	**Yes**	class Cephalopoda
	No	class Gastropoda

CLASSIFYING CELLS

NAME:

DATE:

Identify whether the cells or organisms shown below are prokaryotic or eukaryotic.

- Differentiate between types of cells.

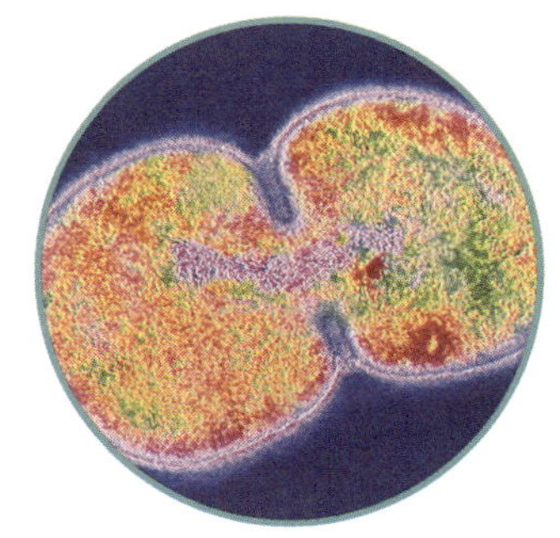

1. prokaryotic

2. eukaryotic

3. eukaryotic

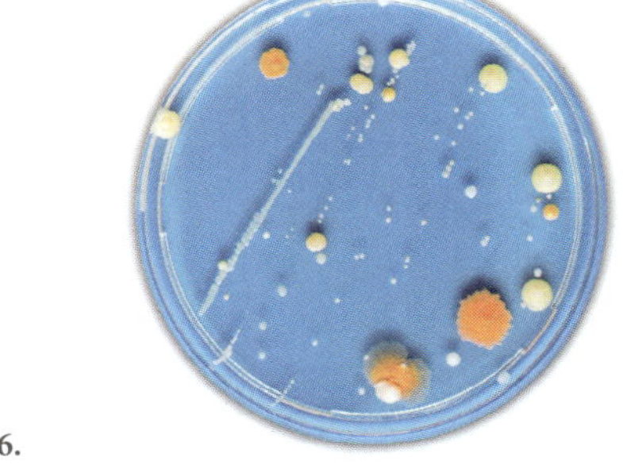

4. eukaryotic

5. eukaryotic

6. prokaryotic

NOTES

IDENTIFYING TYPICAL CELL PARTS

NAME:

DATE:

- List cellular organelles with their functions.

Each of Questions 1–8 below describes a cellular structure. In each blank, write the name of the structure that is being described. Then draw the structure in the cell diagram. Use the number of the question to identify the structure. The cell membrane has been done for you as an example.

1. boundary of the cell

   cell membrane

2. a jelly-like fluid that surrounds organelles

   cytosol

3. modifies and packages proteins into vesicles

   Golgi apparatus

4. holds the DNA inside a membrane

   nucleus

5. produces and transports proteins and other molecules

   endoplasmic reticulum

6. builds proteins

   ribosome

7. contains enzymes for breaking down cellular waste products

   lysosome

8. manufactures most of the cell's ATP

   mitochondrion

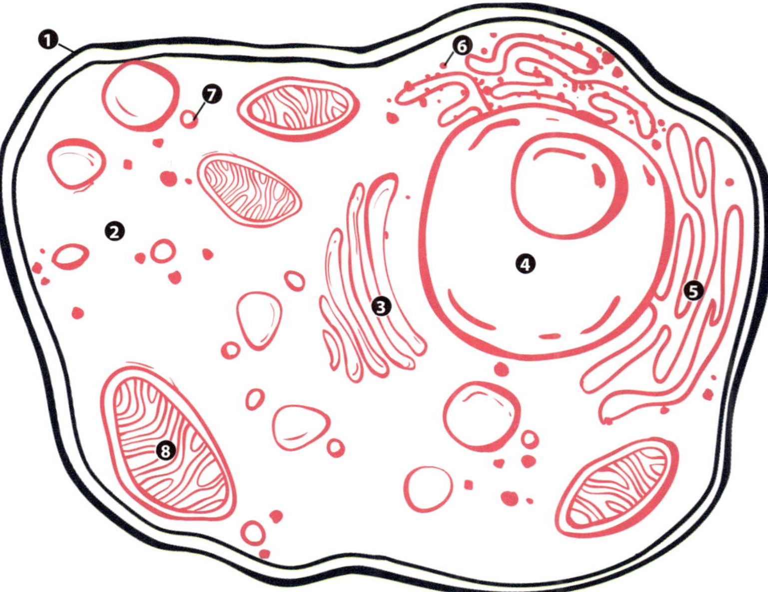

9. Is this a bacterial cell or an animal cell?

   an animal cell

10. How can you tell?

    This cell has a nucleus, so it is a eukaryotic cell. Bacterial cells are prokaryotic.

NOTES

REVIEW 2C

EXPLAINING PHOTOSYNTHESIS

NAME:

DATE:

Below are several groups of terms. In each group, three of the four terms are most closely related to one another. Draw a line through the least related term, and then write a sentence using the remaining three terms. Use your sentence to show how the terms are related. You may change the form of the term in your sentence (for example, *product* to *products* or *stored* to *storage*).

1. photosynthesis / ~~cellulose~~ / sugar / oxygen

   Answers will vary. Example: Sugar and oxygen are products of photosynthesis.

2. light energy / light-dependent phase / photosynthesis / ~~sound energy~~

   Answers will vary. Example: Light energy is absorbed during the light-dependent phase of photosynthesis.

3. ~~consumer~~ / photosynthesis / chlorophyll / pigment

   Answers will vary. Example: Chlorophyll is a pigment used in photosynthesis.

4. chlorophyll / chloroplasts / ~~oxygen~~ / organelles

   Answers will vary. Example: Chlorophyll is found in organelles called chloroplasts.

5. ~~sugar~~ / light / carbon dioxide / water

   Answers will vary. Example: Carbon dioxide, light, and water are starting materials for photosynthesis.

6. sugar / oxygen / ~~pigment~~ / products

   Answers will vary. Example: Sugar and oxygen are products of photosynthesis.

7. ~~sugar~~ / electrons / protons / water

   Answers will vary. Example: Water is split to produce electrons and protons.

8. producers / food / photosynthesis / ~~consumers~~

   Answers will vary. Example: Producers make their own food through photosynthesis.

REVIEW 2C OBJECTIVE

- Summarize how cells obtain, store, and use energy.

EXPECT VARIETY 

The answers given are intended only as suggestions. Student answers will vary quite a bit. Grade for accuracy, not for conformity to the suggested answer.

EXPLAINING PHOTOSYNTHESIS **23**

NOTES

9. <s>sugar</s> / light / chlorophyll / absorb

Answers will vary. Example: Chlorophyll absorbs light energy.

10. chlorophyll / energy / <s>light-independent</s> / water

Answers will vary. Example: Chlorophyll uses energy from the sun to split water molecules.

LAB 2D

THE NOSE KNOWS

Diffusion Rates

WHAT CONDITIONS CAN SPEED UP DIFFUSION?

Mmmm … pizza for dinner. How can you tell when your room is at one end of the house and the kitchen is at the other? Diffusion!

Here's how it works. Smells are actually small particles that are constantly in motion. If a certain type of particle is more concentrated in one place (like the aroma of pizza in the kitchen), the molecules begin spreading out (reaching into your room). Then those pizza particles travel up your nose and ring the dinner bell in your brain!

You've already learned about diffusion—the movement of particles in a cell. In this lab activity, you will investigate how stirring and temperature affect the speed of diffusion.

Procedure

STIRRING AND DIFFUSION

1. Write a hypothesis stating whether you think diffusion will occur more quickly in stirred or unstirred water.

   See TE margin for answer.

2. What is the independent variable in your experiment? What is the dependent variable?

   The independent variable is whether the beaker is stirred. The

   dependent variable is the rate of diffusion.

NAME:

DATE:

Key Question

- Is the rate of diffusion affected by stirring or temperature?

Equipment

beakers, 250 mL (2)

water (hot and cold)

dark food coloring

measuring spoons

stirring rod or spoon

stopwatch or a clock with a second hand

goggles

LAB 2D OBJECTIVES

- Compare different rates of diffusion.

- Determine how different factors affect diffusion rates.

GOGGLES 

As a general rule of thumb, scientists use goggles whenever chemicals or glassware is being used. Other hazardous conditions may also warrant the use of goggles.

COMPARING RESULTS 

You might have your students compare their results with those of other students. Or you could have each student repeat this experiment once or twice and take an average of the data.

Question 1 Answer

Answers will vary. Check that the student has predicted that one of the beakers will experience faster diffusion or that the rates will be the same.

NOTES

If you have access to a hot plate with a magnetic stirrer, place a magnet stir bar in the beaker and turn the magnetic stirrer on low instead of having a student stir the water with a spoon.

BE READY! 

Make sure that students are ready to promptly start timing since the stirred diffusion will proceed quickly.

Question 3 Answer

Answers will vary. Check that the student has predicted that one of the beakers will experience faster diffusion or that the rates will be the same.

❶ Fill two clean glass beakers each with 200 mL of room-temperature water.

❷ Place the two beakers next to each other and wait for the water to become still.

❸ Measure two identical amounts of dark food coloring (about 2 drops). Using a stirring rod or a spoon, stir the water in one beaker. Then immediately pour one portion of food coloring into the beaker of stirred water and the other portion into the beaker of unstirred water. Be sure to pour both portions of food coloring at the same time.

❹ Continue stirring the one beaker and use a stopwatch to measure the time required for each beaker to complete diffusion. Diffusion is complete when all the water in the beaker is the same color; no streaks or currents will be visible. In Table 1, record the time it took each beaker to complete diffusion in to the nearest tenth of a second. If ten minutes pass without diffusion being complete in one or both beakers, write "> 600" in Table 1.

❺ Dispose of the water and food coloring and rinse out the beakers.

TEMPERATURE AND DIFFUSION

3. Write a hypothesis stating whether you think diffusion will be faster in hot or cold water.

   See TE margin for answer.

4. What is the independent variable in your experiment? What is the dependent variable?

   The independent variable is the temperature of the water. The dependent variable is the rate of diffusion.

❶ Set up two beakers of water as you did in the previous procedure. This time, use very hot water in one beaker and very cold water in the other.

NOTES

② Prepare two identical amounts of dark food coloring as before. Do not stir the water in either beaker. Pour equal portions of food coloring into each beaker and start timing. Be sure to pour both portions of food coloring at the same time.

③ Use a stopwatch to measure the time required for each beaker to complete diffusion. In Table 2, record the time for each beaker to the nearest tenth of a second. If ten minutes pass without diffusion being complete in one or both beakers, write "> 600" in Table 2.

5. According to the data that you recorded in Table 1, was your hypothesis about stirring correct?

Answers will vary.

6. According to the data that you recorded in Table 2, was your hypothesis about hot and cold water correct?

Answers will vary.

CONCLUSION

7. Explain why diffusion happened faster in one beaker in each experiment.

Both stirring and heating increased the movement of the molecules in the beaker. The increased movement causes diffusion to happen more quickly.

NOTES

8. After you poured food coloring into a beaker, was there a concentration gradient in the beaker? Explain.

Yes. The food coloring was more concentrated in the region of the beaker where the food coloring was poured.

9. After the experiment was completed, was there a concentration gradient in the beaker? Explain.

No. The food coloring was evenly spread throughout the beaker.

10. In cells, diffusion is an example of what kind of transport?

passive transport

11. What is the major difference between diffusion in your experiment and diffusion in cells?

Diffusion in the cell occurs across a membrane.

Going Further

12. How could you apply what you have learned about diffusion to cooking, treating wastewater, or even deciding where you put an air freshener in your house?

Answers will vary, but students should recognize that stirring the medium and heating the medium will speed mixing through diffusion. That could apply to making a gallon of sweet tea, mixing chemicals to treat water, or making air smell fresher.

Table 1

Diffusion Rate with Stirring

Beaker	Time (s)
stirred	117.5
unstirred	> 600

Table 2

Diffusion Rate with Temperature

Beaker	Time (s)
cold	> 600
hot	66.3

SAMPLE DATA

The data in Table 1 and Table 2 are sample data. Your students' data may be different.

NOTES

LAB 2E

PLANTS IN THE DARK

Starch and Photosynthesis

DOES A PLANT USE PHOTOSYNTHESIS TO FORM STARCH?

You've been learning about the process of photosynthesis. Cells in plant leaves use photosynthesis to create molecules of a sugar called *glucose*. Some of the glucose is used right away by the plant cells. The rest of the glucose that plants don't immediately use is made into larger sugar molecules called *starch*. Some of the starch molecules are stored in storage organs such as fruit and roots. Other starch molecules are kept in the leaf cells.

But how does a plant get energy during the night? It uses its stored starch for energy. But can't it also continue to use photosynthesis to make starch in the dark? In this lab activity, you will test whether a plant can actually make starch when it does not have access to light.

Procedure

SETUP

1. Place a healthy geranium in a dark area for two days (a cabinet or cupboard works well).

2. Cut 3 pieces of aluminum foil into rectangles that are each large enough to cover half a leaf.

3. Use paper clips to attach the aluminum foil pieces to the upper surface of three leaves that are still attached to the plant, as shown at right. Be sure that the foil edges adhere closely to each leaf or even bend down slightly over the edges of the leaf. Remember to leave half the leaf exposed.

4. Now place the plant in a sunny window or another well-lit area for one or two days.

1. Write a hypothesis about how the levels of starch in the covered and uncovered parts of the leaf will differ.

   Answers will vary.

2. What is the independent variable in your experiment? What is the dependent variable?

   The independent variable is exposure to light. The

   dependent variable is the presence of starch.

NAME:

DATE:

Key Questions 

- How does the presence of starch show that photosynthesis has occurred?

- Is light necessary for photosynthesis?

Equipment 

geranium or coleus plant

aluminum foil

paper clips

alcohol (70% ethanol or 70% isopropyl), 100mL

beaker, 400 mL

hot plate or other heat source (not an open flame)

water

petri dish

iodine solution

goggles

- Test for the presence of starch.

- Demonstrate that light is necessary for photosynthesis.

PLANT LEAVES 

Other kinds of plants besides those included here will work as well. The important thing is that the plant has broad leaves, preferably 2 in. wide or greater.

PREPARING FOR THE LAB ACTIVITY 

To save class time, you might want to perform the lab setup over the few days before lab day. The students will then have only to test the leaves on lab day.

PLANTS IN THE DARK **29**

NOTES

HEATING THE ALCOHOL 

The alcohol should be heated on a hot plate or a similar heating appliance. **Do not** heat it with a Bunsen burner or any other equipment that has an open flame. Alcohol vapors could easily be ignited by an open flame.

IODINE SOLUTION 

Iodine solution is a skin and eye irritant. Wear goggles to protect against eye contact. If students get it on their skin, they should wash with soap and water immediately. Iodine solution can also stain clothing.

These leaves have been stained with iodine. The leaf above contains starch. The one below does not.

DATA COLLECTION

The strong green color of chlorophyll can make it hard to see the results of the test for starch. You will use alcohol to remove the chlorophyll from the leaves.

1. Heat the alcohol in a beaker until it is hot but not boiling. Remove from the heat source. You should use enough alcohol to cover the leaves that will be placed in the beaker.

2. Remove the leaves with aluminum foil from the plant. Remove the aluminum foil.

3. Immerse the leaves in the hot alcohol for about five minutes. During this time, the alcohol will remove the chlorophyll from the leaves.

4. Rinse the leaves with water and place each leaf flat in a petri dish.

5. To test for the presence of starch, cover each leaf in the petri dish with iodine solution for five minutes.

6. Rinse off the iodine solution and examine the leaves. Iodine stains starch a dark blue.

7. Complete Table 1 on the basis of your observations. Indicate what color iodine turned the leaves and whether starch was present by comparing your leaf to the ones at left.

3. Did starch form in the covered or uncovered part of the leaf?

the uncovered part

CONCLUSIONS

4. What does the presence of starch indicate?

Photosynthesis has occurred.

5. What does the absence of starch indicate?

Photosynthesis has *not* occurred.

6. Why did you have to put the plant in a dark area for two days?

When the plant is stored in a dark area, cellular respiration uses up any starch already stored in the leaves, and the plant is unable to replace it by photosynthesis.

7. What would you have discovered if you had tested for starch immediately after bringing the plant out from the dark? Explain.

The plant would have had no starch after two days with no photosynthesis, so no starch would have been indicated.

NOTES

Going Further

8. How do consumers (like you) use starch?

Consumers break starch into glucose molecules, which are then
used to produce energy through cellular respiration.

____

Table 1

Presence of Starch in Leaves

Leaf	Leaf Section	Color	Starch Present?
leaf 1	covered	brown	N
	uncovered	dark blue	Y
leaf 2	covered	brown	N
	uncovered	dark blue	Y
leaf 3	covered	brown	N
	uncovered	dark blue	Y

NOTES

____

____

____

____

____

THE STRUCTURE OF DNA

NAME:

DATE:

You've been learning that a DNA molecule exists in the form of a double helix. Each half of the helix is a series of nucleotides. The DNA strand shown below is missing half of its nucleotides. You get to make the missing half! Complete the key to the right, then use it to create the missing half of the DNA strand.

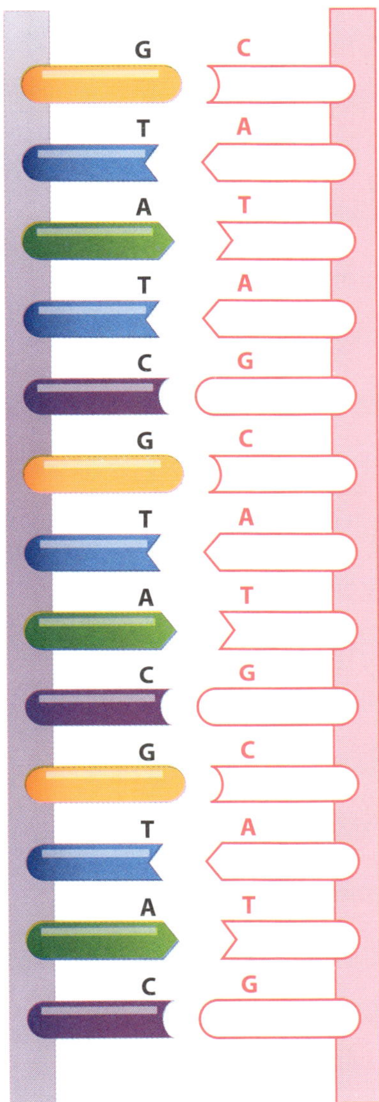

KEY

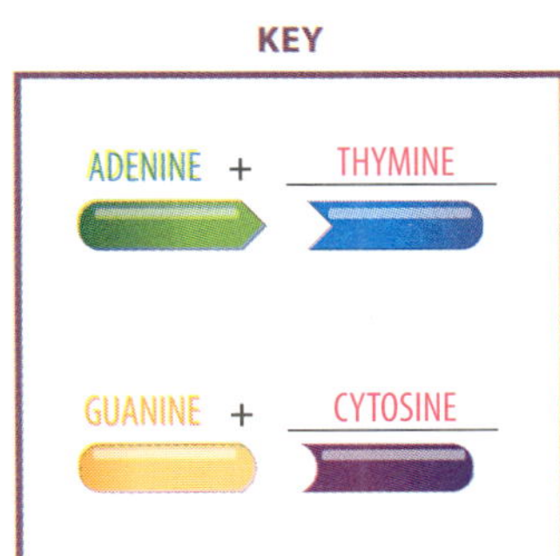

REVIEW 3A OBJECTIVE

- Describe the structure of DNA.

MAKE IT A MODEL 

With a bit of planning ahead, you can modify this review activity into an opportunity for hands-on modeling. You will need some kind of object that comes in four different colors to serve as nucleotides (e.g., pipe cleaners, colored marshmallows) and something to serve as sides and rungs of the DNA molecule (e.g., paperclips, toothpicks). Have students use the side shown here as a template for one side of their DNA molecule, then complete the molecule with the appropriate pieces.

NOTES

REVIEW 3B

TRANSCRIPTION OR TRANSLATION?

NAME:

DATE:

Identify whether each of the following processes happens during transcription (TS) or translation (TL).

REVIEW 3B OBJECTIVE

- Summarize the processes of transcription and translation.

TS	**1.**	Happens at ❶.
TL	**2.**	Happens at ❷.
TS	**3.**	The two sides of a DNA molecule are temporarily unzipped by enzymes.
TL	**4.**	Anticodons are matched to codons.
TS or TL	**5.**	Adenine is paired with uracil instead of thymine.
TL	**6.**	A protein molecule is produced.
TL	**7.**	The information in a strand of RNA is "read" by a ribosome.
TL	**8.**	Amino acids are placed in a sequence.
TS	**9.**	A strand of messenger RNA is made.

NOTES

REVIEW 3C

THE CELL CYCLE

NAME:

DATE:

Table 1 shows a series of diagrams illustrating each phase of the cell cycle. Beside the diagrams is a column for the name of the phase being illustrated and another column for the description of what is happening in that phase. Fill in the missing names or descriptions. In your descriptions be sure to use the following words if they apply: *chromosomes, sister chromatids,* and *spindle.*

Table 1

Cell Diagram	Cell Division Phase	Description
	interphase	Genetic material is duplicated.
	prophase	First stage of mitosis: The nuclear membrane disappears and the chromosomes coil up. The spindle begins to form.
	metaphase	Second stage of mitosis: Chromosomes composed of sister chromatids line up at the center of the spindle.
	anaphase	Third stage of mitosis: Each pair of sister chromatids separates into two daughter chromosomes, which begin to migrate to opposite ends of the cell.
	telophase	Final stage of mitosis: The daughter chromosomes reach the ends of the spindle and begin to uncoil. New nuclear membranes form around each group of chromosomes.
	cytokinesis	The cytoplasm divides, forming two daughter cells.

REVIEW 3C OBJECTIVE

• Summarize the cell cycle.

INDISTINCT STAGES 

Be sure that students understand that there is not a rigid division between each stage of the cell cycle. For example, cytokinesis may begin before telophase is completed.

NOTES

LAB 3D

DIVIDE AND CONQUER

The Phases of Mitosis

WHAT DO THE PHASES OF MITOSIS LOOK LIKE?

You've been looking at some illustrations of the different phases of the cell cycle, but what do they really look like? One of the places where cells rapidly divide is in the root tips of plants. Let's take a look at some root tip cells to see what the phases of mitosis look like in real cells.

Procedure

1. What kind of plant are you observing?

   Answers will vary.

❶ Focus your microscope on one of the root tips on the prepared slide.

❷ Locate the area just above the root cap. (See image at right.)

In this area, mitosis occurs rapidly to form new root cells so that the root can grow. When this root tip was prepared for microscope viewing, many of the cells in this area were in various phases of mitosis. The root tip was then treated with special stains to allow you to see the chromosomes.

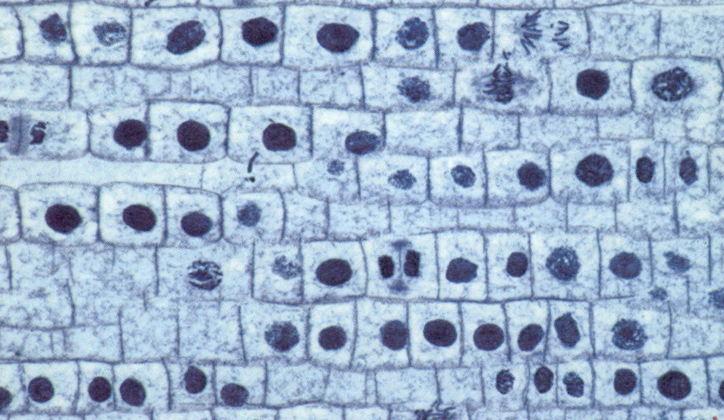

❸ So that you can see the chromosomes in the various phases of mitosis, focus your microscope on high power.

❹ Scan your slide by slowly moving the slide back and forth. Look for cells in the various phases of mitosis.

❺ Find what you think is a good example of one of the phases of mitosis. Position that cell in the center of your microscope's viewing field.

2. What phase of mitosis do you believe this is?

   Answers will vary.

NAME:

DATE:

Key Questions 

• What are the different stages of mitosis?

• How can I identify these stages in images of real cells?

Equipment 

microscope

prepared slides of onion, hyacinth, or lily root tips showing the phases of mitosis

LAB 3D OBJECTIVES

• Identify the phases of mitosis in plant cells.

• Describe the sequence of phases in mitosis.

SAVING TIME WITH MICROSCOPE SETUP 

You can shorten this exercise by finding the phases of mitosis in the microscopes before class begins. Arrange the phases in order, then have the students view them and answer the questions. This method is also good if you do not have enough microscopes for all the students. To do this exercise as written, you will need to have about one microscope for every two to four students in your class.

HOW SLIDES ARE CREATED 

Explain to your students that slide manufacturers slice the root tips when they prepare the slides. Sometimes the slice is made close to the edge of the cell, making the cell appear empty. Actually, the chromosomes that belong in that empty cell are on another slide.

SETTING UP MICROSCOPE STATIONS

Establish the number of prophases, metaphases, anaphases, and telophases you want your class to find; base that amount on the number of microscopes you have. Try to have an equal number of each phase so that students can conveniently view each phase. List the phases on the whiteboard with a certain number of lines under them. Tell students to write their names on the proper line when they have had their phase approved. Once the lines are filled, no more of that phase should be found. This will then permit those who have not yet found a phase to know which ones to look for.

NOTES

EVALUATING STUDENT RESPONSES 

The descriptions given here are really all that students will be able to identify under a light microscope. They may be able to detect the X-shaped sister chromatids in prophase and metaphase. The spindle is not visible.

NO CYTOKINESIS? 

When observing plant cell division, the students *may* see cell plate formation; they *will not* see cytokinesis.

6. Ask your teacher to check your cell to be sure that it shows a good example of the phase you think you are viewing.

7. After your teacher has approved all the slides, move from one microscope to another, viewing the various phases of mitosis in order. Carefully note the sequence of the positions of the chromosomes in mitosis.

Summing Up

Write a brief description of the positions of chromosomes during mitosis as you saw them in the various microscopes.

3. Prophase:

   In prophase, the chromosomes are coiled up.

4. Metaphase:

   In metaphase, the chromosomes are lined up along the middle of the spindle.

5. Anaphase:

   In anaphase, the chromosomes have divided and the daughter chromosomes are moving away from each other on the spindle.

6. Telophase:

   In telophase, the chromosomes are uncoiled at the ends of the spindle.

7. What other observations about mitosis did you make while looking at these cells?

   Answers will vary. Students may notice that daughter cells are about half the size of regular cells. They may notice the division plate forming between the daughter cells. They may also notice that cells in interphase (technically a stage of the cell cycle, not a phase of mitosis) have fuzzy-looking nuclei.

NOTES

SIMPLE GENETICS PROBLEMS

NAME:

DATE:

Baa, Baa, Black Sheep. But aren't most sheep white? Sheep can be either white or black. Wool color in some breeds of sheep is controlled by a simple dominant-recessive inheritance pattern. An uppercase *W* represents the dominant allele that causes white wool. A lowercase *w* represents the recessive allele. In the absence of any dominant allele, this allele causes sheep to have black wool.

1 Using the information above, complete Table 1.

Table 1

Genotypes and Phenotypes for Color in Sheep

Genotype	Dominant or Recessive Trait Expressed?	Purebred or Hybrid?	Color of Wool
WW	dominant	purebred	white
ww	recessive	purebred	black
Ww	dominant	hybrid	white

2 Two white sheep with genotypes *Ww* have offspring. Create a Punnett square for the cross using the blank Punnett square below.

	W	*w*
W	*WW*	*Ww*
w	*Ww*	*ww*

1. What color of wool will a lamb from the cross in Step **2** likely have? Explain.

The lamb will likely be white since each offspring has a 3-in-4 chance of being white.

2. A white sheep having the genotype *WW* is crossed with a white sheep with the genotype *Ww*. What color of wool will their offspring have?

white

3. A black male sheep mates with a white female sheep. The female sheep has twins. One is black and the other is white. What is the genotype of the white female sheep? Explain.

Ww. If her genotype was *WW*, all the offspring would have been white.

REVIEW 4A OBJECTIVE

- Predict the outcome of a cross with a simple dominant-recessive inheritance pattern using Punnett squares.

COLLABORATIVE ACTIVITIES 

Some students may struggle with Reviews 4A and 4B and may find it helpful to do them as part of a collaborative activity.

QUESTION 3 

Question 3 requires the student to reason from known information to unknown information. This reasoning process may be too advanced for some seventh graders. Be prepared to help students work through this question.

NOTES

Rabbits come in a variety of colors. In some breeds, the color of a rabbit's coat is controlled by a single gene. The dominant trait, represented as R, is brown fur. The recessive trait, r, is white fur. Using this information, complete Table 2.

Table 2

Genotypes and Phenotypes of Rabbits

Genotype	Dominant or Recessive Trait Expressed?	Purebred or Hybrid?	Brown or White?
RR	dominant	purebred	brown
rr	recessive	purebred	white
Rr	dominant	hybrid	brown

	R	R
R	RR	RR
r	Rr	Rr

3 Complete the Punnett square on the left.

4. What is the phenotype of the parents?

They are both brown.

5. What would the offspring look like?

They will all be brown.

6. Which of the following is true of the offspring?

   a. They would all be purebred for brown fur.

   b. They would all be purebred for white fur.

   c. They would all be hybrids with brown fur.

   d. Some would be purebred for brown fur, and some would be hybrids with brown fur.

   e. Some would be purebred for brown fur, and some would be hybrids with white fur.

d

NON-MENDELIAN PUNNETT SQUARES

NAME:

DATE:

Punnett squares can be used to represent simple genetic crosses for characters that do not have a simple dominance pattern. Fill in the Punnett squares below to answer the questions that follow each one.

The Punnett square below represents a cross between two purple radishes. Radish color is an incomplete dominant trait. Purebred radishes are either red or white.

	C^r	C^w
C^r	C^rC^r	C^rC^w
C^w	C^rC^w	C^wC^w

1. What radish color(s) will be produced by this cross?

   red, purple, and white

2. A farmer performed this cross and grew 96 radishes from the seeds. How many purple offspring can reasonably be expected from this cross?

   a. 25

   b. 50

   c. 21

   b

REVIEW 4B OBJECTIVE

- Predict the outcomes of crosses with non-Mendelian inheritance patterns.

NOTES

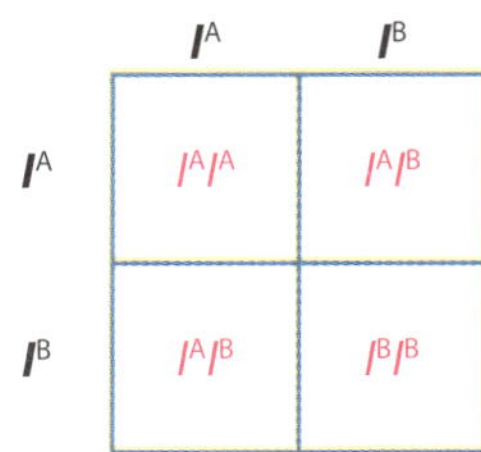

Review the inheritance of blood types on page 76 of your textbook. A married couple both have type AB blood as shown in the Punnett square above.

3. What blood types might their children have?

   a. type A

   b. type B

   c. type AB

   d. type O

   a, b, c

4. If this couple had seven children, which blood type would you expect to be the most common? Why?

   Type AB would probably be most common since each child has a 50% chance of having type AB but only a 25% chance each of being type A or type B.

REVIEW 4C

VARIATION FROM SEXUAL REPRODUCTION

NAME:

DATE:

Sexual reproduction offers one clear advantage that asexual reproduction (or cloning) does not offer—variation. Organisms produced by sexual reproduction are similar to, but not exactly like, their parents. In this review you are going to demonstrate that animals produced by sexual reproduction can have certain characteristics that are different from those of their parents.

REVIEW 4C OBJECTIVE

- Predict the results of changes in a population.

1. What is the fur pattern of the mother shown below?

   Answers will vary. Example: white fur

2. What is the fur pattern of the father?

   Answers will vary. Example: black hair with light-colored chest

3. What is the fur pattern of offspring ❶?

   Answers will vary. Example: light-brown fur with some mixed-dark fur

4. What is the fur pattern of offspring ❷?

   Answers will vary. Example: dark fur with some lighter fur mixed in

5. What is the fur pattern of offspring ❸?

   Answers will vary. Example: light-brown fur with white chest and muzzle, with a white stripe down the face

NOTES

6. What is the fur pattern of offspring **4**?

   Answers will vary. Example: dark fur with white chest and white patch on head

7. How many of the offspring had fur color and patterns identical to the mother's or father's?

   None of them did.

8. How many different kinds of fur coat patterns were found in the litter?

   Accept either four or five.

9. How many different kinds of fur coat patterns were demonstrated by the parents?

   two

10. If the offspring were genetically identical to their mother, would the color of their fur coats be different from hers?

    No.

11. Other than fur coat pattern, list five physical traits in which wild animals could vary from their parents.

    Answers will vary but may include size, fur color, temperament, ear size, fur/hair length, and eye color.

12. How would the ability to produce different fur patterns help a group of animals to survive for thousands of years?

    Answers will vary. Certain patterns might be a better camouflage in different areas or at different times of the year.

13. Why would *not* having variation be harmful to a species?

    If every individual of a species were exactly the same, a natural event (predator, disease, climate, etc.) that could eliminate one due to a particular inherited trait could eliminate all.

PUPPY COAT PATTERNS [!?!]

The answer to Question 10 is based on the simplified presentation of genetics in Chapter 4. In reality, it is possible for two dogs to be genetically identical for a coat pattern and yet express that pattern differently. For example, this is true with variegated coat patterns such as Blenheim or brindle. A particular combination of genes determines the pattern, but the exact way the pattern is expressed may be random or influenced by the environment.

THE AMAZING SPUDOODLE

Inheritance of Traits

NAME:

DATE:

HOW DO AN ORGANISM'S TRAITS REFLECT ITS PARENTS' TRAITS?

This strange-looking thing is a spudoodle. Like all organisms, its traits are determined by its genes. And it inherits its genes from its parents. The spudoodle reproduces sexually, so one-half of its 16 chromosomes comes from its father and the other half comes from its mother. Thus, the spudoodle has 8 chromosome pairs. You will be making your own baby spudoodle according to its parents' traits.

Procedure

THE PARENTS' GENOTYPES

See Table 1 for the spudoodle's inherited traits. You will use a coin toss to create the genotypes of its parents.

1. For each of the father's traits, flip a coin twice. A flip of heads represents the dominant allele. A flip of tails represents a recessive allele. Record the father's genotype in the *Father's Genotype* column of Table 1.

2. Repeat Step 1 to determine the genotype of your spudoodle's mother. Record the mother's genotype in the *Mother's Genotype* column.

YOUR SPUDOODLE'S GENOTYPE AND PHENOTYPE

Your spudoodle will receive one allele from each of its parents for each gene.

1. If your spudoodle's father is purebred for a gene, your spudoodle will receive that allele. Record that allele in the *My Spudoodle's Genotype* column of Table 1.

2. But if your spudoodle's father is hybrid for a gene, flip a coin. A flip of heads indicates that your spudoodle receives the dominant allele. A flip of tails indicates that your spudoodle receives a recessive allele. Record that allele in the *My Spudoodle's Genotype* column.

3. Repeat Steps 1 and 2 for your spudoodle's mother and record the alleles in the same *My Spudoodle's Genotype* column.

Key Questions

- What is the relationship between inheriting chromosomes and inheriting traits?

- How are inherited traits expressed?

Equipment

potato

toothpicks

coins (penny, nickel, and quarter)

popped popcorn or puffed-corn cereal

miniature marshmallows (green, pink, yellow, and white)

pipe cleaners

pushpins or thumbtacks (green, red, white, and yellow)

LAB 4D OBJECTIVES

- Demonstrate the relationship between the inheritance of chromosomes and the inheritance of traits.

- Model the expression of dominant, recessive, and incompletely dominant traits.

PREPARATION 

You may want to have students bring in their own potatoes and coins in a plastic bag with their name on it.

Cut toothpicks in half for use in this lab. Full-size toothpicks stick out so far that the spudoodle becomes unwieldy. Half toothpicks work better.

Many substitutions are possible in this lab activity. If different colors of miniature marshmallows are unavailable, use highlighters or food coloring to color some.

SEQUENCE OF BUILDING THE SPUDOODLE 

It is strongly recommended that the students add the structures in the order listed in Table 1. This is especially true with the mouth and the legs. Inserting the coin for the mouth requires a decent bit of force, and inserting the toothpicks for the dorsal spines would make holding the potato difficult. The legs will start to become loose if added too early as the potato is jostled when the other structures are pushed into it.

NOTES

④ Using the genotype for each trait, determine your spudoodle's phenotype for each trait. Record your data in the *My Spudoodle's Phenotype* column.

⑤ Have a classmate check to see that your spudoodle's recorded phenotypes are consistent with your recorded genotypes, putting a check mark for each trait in the *Classmate Check* column of the table. Have your classmate sign at the bottom.

BUILD YOUR SPUDOODLE

① Assemble your baby spudoodle according to the traits coded for by its chromosomes. Use toothpicks to attach marshmallows, popcorn, and puffed-corn cereal. Look at the spudoodle in the photo to see where the various body parts belong. You'll likely want to put the mouth on first and the legs last. Use the following materials for your spudoodle's traits.

- Hair—toothpicks (HH or Hh) or none (hh)
- Mouth—coins (quarter [AA], nickel [Aa], or penny [aa])
- Ears—popcorn (EE or Ee) or puffed-corn cereal (ee)
- Eyes—green (BB or Bb) or white (bb) marshmallows
- Dorsal spines—yellow (DD), pink (Dd), or white (dd) marshmallows
- Tail—curly (RR or Rr) or straight (rr) pipe cleaners
- Legs—green (GG), red (Gg), or white (gg) thumbtacks
- Nose—two (NN or Nn) yellow thumbtacks or one (nn) yellow thumbtack

② When you are finished, name your spudoodle and place it in the spudoodle pen designated by your teacher.

1. What is your spudoodle's name?

Answers will vary.

2. List the spudoodle body parts that have traits controlled by simple dominant-recessive genes.

hair, ears, eyes, tail, and nose

3. List the spudoodle body parts that have traits controlled by incompletely dominant genes.

mouth, dorsal spines, and legs

4. Does the inheritance of a large mouth have any effect on spudoodle eye color? Why or why not?

No. Eye color traits are controlled by a different chromosome

pair.

5. Does the sex of the parent influence whether a trait is more likely to be inherited?

No.

NOTES

6. Is it possible for a spudoodle with yellow dorsal spines to have a baby spudoodle with white dorsal spines? Why or why not?

No. If the parent had yellow dorsal spines, it would have two alleles for the yellow trait (*DD*) and would thus be unable to pass on the white allele (*d*) to its offspring. The offspring must get a white allele (*d*) from each of its parents to have white dorsal spines.

Going Further

7. How does your spudoodle model what happens in the genetics of a real organism?

Organisms have multiple traits that are governed by different types of inheritance.

Table 1

Possible Genotypes	Possible Phenotypes	Father's Genotype	Mother's Genotype	My Spudoodle's Genotype	My Spudoodle's Phenotype	Classmate Check
HH or *Hh*	hair					
hh	no hair					
AA	large mouth					
Aa	medium mouth					
aa	small mouth					
EE or *Ee*	popcorn ears					
ee	puffed-corn ears					
BB or *Bb*	green eyes					
bb	white eyes					
DD	yellow dorsal spines					
Dd	pink dorsal spines					
dd	white dorsal spines					
RR or *Rr*	curly tail					
rr	straight tail					
GG	green legs					
Gg	red legs					
gg	white legs					
NN or *Nn*	two-nostril nose					
nn	one-nostril nose					

Signature of classmate who checked your spudoodle: ________________________

NOTES

LAB 4E

FADING COLORS

Modeling Natural Selection

HOW DOES NATURAL SELECTION AFFECT DIFFERENT POPULATIONS?

When animals from each kind left the Ark, they began to reproduce. They then started to spread across the world. At first, each population of a kind would have probably looked similar to each other. But as natural selection affected the different populations, certain traits (e.g., spotted fur) became less common in some populations and more common in others. Eventually, the populations began to look and behave differently. God has built variety into living things to allow them to adapt to different environments.

In this lab activity, you will model how this may have happened. Remember that a model is a representation of a complex structure or process. This model is a simplified picture of how different species of organisms could have come from the original animals that survived the Flood.

NAME:

DATE:

Key Questions 

- Is it possible to model natural selection?

- How does natural selection change populations over time?

Equipment

colored marbles (10 each of 6 different colors)

empty box

game die

LAB 4E OBJECTIVES

- Use a model to represent the effects of natural selection.

- Explain how a population changes due to natural selection according to this model.

PREPARATION 

This lab activity works well as a class activity. Consider dividing the class into four groups, each one with a separate population. As you roll the die, the students can remove the marbles from their populations.

Marble sets for Chinese Checkers work well for this activity.

NOTES

CALCULATING PERCENTAGES

Students with less experience in math may require assistance with calculating the percentage of each color in each population. For all students, a calculator or calculator app will facilitate the process.

CORRECT DENOMINATOR 

Be sure to remind students that when calculating the percentage of the ending population in Step 8, the denominator should be the number of marbles left in the population. Some students may mistakenly use 15 as the denominator since that was the denominator at the beginning of the activity.

Procedure

❶ Assign each marble color a number by filling out the *Colors* row of Table 1.

1. What does each color marble represent?

The color of each marble represents a trait of a certain character.

❷ Mix the 60 marbles in the box and then randomly divide them into 4 different groups. Put them into a cup or beaker to contain them.

2. What do these groups of marbles represent?

They represent four separate populations.

For Step ❸ below, you will calculate the percentage of each color in Population 1 by dividing the number of each color of marble by 15 and multiplying by 100%.

3. Why do you divide the number of each color marble by 15?

Each population has 15 marbles.

❸ Record in Table 1 the percentage of each color in Population 1 at the beginning of the experiment in the *Population 1, beginning* cell.

❹ Repeat Step ❸ for the other three populations.

❺ Roll the game die for Population 1. Find the color that corresponds to that number and remove all the marbles of that color from Population 1. Set aside the marbles you removed.

4. What does rolling the die represent in our model of variety within a kind of organism?

Rolling the die represents an environmental pressure against

a certain trait in a population.

❻ Repeat Step ❺ for the other three populations.

❼ Repeat Steps ❺ and ❻ five additional times.

5. What do you notice about the number of marbles after each die roll?

The number of marbles usually goes down.

6. What do you notice about the number of the colors of marbles after each die roll?

There is usually less variety in color.

❽ Record in Table 1 the percentage of each color in each population at the end of the experiment in the appropriate *Population, end* rows.

NOTES

7. At the end of the experiment, how is Population 1 different from Population 2?

<u>Answers will vary. Student should indicate which colors are</u>

<u>present in one population but absent in the other.</u>

____

8. Did any populations gain any traits?

<u>No.</u>

9. How does this model the action of natural selection in a real population?

<u>Natural selection can only remove alleles from a population;</u>

<u>it cannot add them.</u>

Going Further

10. Given what you know about natural selection, does variety in a population increase or decrease over time? Explain.

<u>Variety in a population decreases over time as natural selection</u>

<u>removes traits.</u>

11. Compare the genetic diversity of the population of the dog kind twenty-five years after the Flood with a population of foxes in Tennessee today.

<u>The genetic diversity of the population of the dog kind</u>

<u>twenty-five years after the Flood would have been greater than</u>

<u>the fox population in Tennessee today. Natural selection has</u>

<u>reduced genetic diversity, separating foxes from dogs, wolves,</u>

<u>jackals, and other dog species.</u>

SAMPLE DATA ✓

The data in Table 1 is sample data. Your students' data may be different.

Table 1

Colors		Black	Blue	Green	Red	Yellow	White
Number		1	2	3	4	5	6
Population 1	beginning	0%	13%	13%	27%	27%	20%
	end	0%	100%	0%	0%	0%	0%
Population 2	beginning	33%	20%	20%	7%	7%	13%
	end	56%	0%	33%	11%	0%	0%
Population 3	beginning	13%	13%	13%	13%	20%	27%
	end	0%	0%	40%	0%	60%	0%
Population 4	beginning	20%	20%	20%	20%	13%	7%
	end	50%	0%	0%	50%	0%	0%

NOTES

____

____

____

____

____

____

____

REVIEW 5A

EVIDENCE OR INFERENCE?

NAME:

DATE:

By carefully observing a dead animal's bones, scientists can tell something about the animal's size, its proportions, and a few other things. But they can't easily tell when the animal lived, where it lived, how it interacted with its young, or what other kinds of organisms lived around it. Scientists who study fossils often make an *inference*, or reasonable conclusion, about this kind of information. While this is a legitimate aspect of scientific research, sometimes two scientists can come to completely different conclusions.

Read the following description of the *Tyrannosaurus rex*.

The *Tyrannosaurus rex* was a true monster. Standing up to 6 m tall, it could easily have looked into a second story window. Very few organisms would have been too big to fit into its mouth with a jaw over 1 m long. And anything too large could have been easily torn into bite-sized pieces by its huge teeth and powerful bite. It walked on its powerful back legs and could run up to 70 kph. Although its front legs were very small compared with the rest of its body, they were extremely muscular. They were strong enough to grab the *T. rex*'s prey before the huge teeth delivered the death bite. Its body was covered in a mix of scales and feathers that would have functioned in communicating with the other members of its pack.

In Questions 1–12, indicate whether the description is evidence (E) or an inference (I).

E	**1.**	stood 6 m tall
I	**2.**	had powerful hind legs
I	**3.**	ran 70 kph
E	**4.**	had huge teeth
I	**5.**	was a predator
I	**6.**	had scales and feathers
E	**7.**	had small front legs
I	**8.**	had very muscular front legs
E	**9.**	had a jaw measuring 1 m
I	**10.**	walked on its back legs
I	**11.**	used feathers to communicate
I	**12.**	lived in a pack

EVIDENCE OR INFERENCE? **55**

REVIEW 5A OBJECTIVE

- Explain why worldview affects one's construction of a history of change in life on Earth. **BWS**

DINOSAURS AND THE BIBLE

Most young people love dinosaurs, but much information on dinosaurs is permeated by an evolutionary worldview. Some helpful books and DVDs dealing with dinosaurs from a biblical perspective are available from the Institute for Creation Research and Answers in Genesis.

REAL FACTS ABOUT *TYRANNOSAURUS*

Nearly all statements about *Tyrannosaurus* beyond its physical dimensions are controversial. Paleontologists have hotly debated whether it could run, whether it was a predator or scavenger, and what body coverings it had. Please note that many of the statements given in the paragraph on the *Tyrannosaurus* are inferences and should not be taken at face value. They are presented in this way for the purpose of teaching students to think critically about information. Many museums and popular-level scientific publications will present inferences as fact, even though the inferences may be contested by a significant number of scientists.

EVIDENCE OR INFERENCE?

Students may be surprised by some of the answers in this review. It is important that they realize that inferences should be based on evidence, and some inferences are so obvious that they can be easily mistaken for the evidence itself. For instance, since no one alive has ever seen a *T. rex*, no one knows that they walked on their hind legs. But given the extremely small size of the forelegs and the huge head, it is extremely hard to imagine another way for it to move. Other inferences are less closely based on evidence and are often, as a result, more controversial, even among paleontologists. For instance, scientists have spent decades arguing about whether the *T. rex* was an active predator, a scavenger, or both.

NOTES

13. Study the illustration of the *T. rex* below. List any features that you think are just the artist's own interpretation and that are not strongly supported by physical evidence.

Answers will vary but could include color, skin texture, such as scales, and other features that would not have been preserved in the fossils.

REVIEW 5B

EVOLUTIONISTS' INTERPRETATIONS

NAME:

DATE:

No matter how carefully evolutionists consider the evidence they find, they're trying to fit the pieces into the wrong puzzle. That's because their worldview is wrong.

In the questions below, match each evolutionary statement with the evidence that it is based on by writing the letter of the evidence by the statement.

REVIEW 5B OBJECTIVE

- Explain how evolutionists interpret the evidences for change in living things. **BWS**

EVIDENCE

a. the presence of pharyngeal pouches

b. DNA changed by mutations

c. the presence of animal structures with unknown functions

d. the presence of vertebrate limbs with similar bone structures

e. the presence of DNA in all organisms

f. sediments laid down very slowly in current processes

<u>f</u>   **1.** Earth is millions of years old.

<u>e</u>   **2.** All organisms are descended from a single cell organism.

<u>b</u>   **3.** New traits appeared for natural selection to act on.

<u>c</u>   **4.** Vestigial structures have ceased to function due to evolutionary development.

<u>d</u>   **5.** Frog legs and robin wings are homologous structures showing common ancestry.

<u>a</u>   **6.** Vertebrate embryos show that they all had a common ancestor.

EVOLUTIONISTS' INTERPRETATIONS **57**

NOTES

CREATIONISTS' INTERPRETATIONS

NAME:

DATE:

Creationists must deal with the same evidence that evolutionists deal with. But the biblical worldview leads to different interpretations of that evidence. In Questions 1–6 below, write the creationists' interpretation for each piece of evidence.

1. the presence of pharyngeal pouches

   Answers will vary. Many vertebrates have these structures, but that fact does not show that vertebrates are descended from a common ancestor. Many vertebrates have very different body structures that are controlled by very different genes.

2. DNA changed by mutations

   Answers will vary. Mutations are usually harmful. As time goes on, mutations in a population increase the genetic load of that population.

3. the presence of animal structures with unknown functions

   Answers will vary. The fact that we do not know the function of a structure does not imply that it lacks a function.

4. the presence of vertebrate limbs with similar bone structures

   Answers will vary. Homologous structures could just as easily be interpreted as evidence of a common Designer.

5. the presence of DNA in all organisms

   Answers will vary. God created all organisms with DNA, some of which is unique to each type of organism.

6. sediments laid down very slowly in current processes

   Answers will vary. Sediments were laid down very quickly during the Flood.

REVIEW 5C OBJECTIVE

- Explain how creationists interpret the evidences for change in living things. **BWS**

NOTES

MUTANT PLANTS

Observing Radiation Effects on Seedlings

HOW DO MUTATIONS AFFECT PLANT GROWTH?

For the evolutionary model to be true, new traits must appear in a population. For instance, if color vision appeared in a population of deer, it would be a new trait. Natural selection then would begin working on this trait.

Evolutionists claim that mutations produce these new traits for natural selection to act on. But can mutations really improve an organism's chances of survival?

One cause of mutations is radiation exposure. In this lab activity, you will compare plants grown from seeds exposed to radiation with plants that have not been so exposed. You will also compare the effects of different amounts of radiation exposure on plants.

Procedure

1. Sow between 10 and 20 seeds from the nonirradiated seed packet into Flowerpot #1 (the Control group). Record the number of seeds that you planted in Table 1.

2. Repeat Step 1 for each of the other seed packets, sowing each in a separate flowerpot. Label each pot (#2, #3, #4, and #5) and indicate the amount of irradiation for each in the parentheses of Table 1.

3. Place all the pots in a warm, well-lit area. After two weeks observe the plants and compare their growth. If the plants are just starting to grow, wait another week. Complete the remaining columns of Table 1.

NAME:

DATE:

Key Questions

- Do seeds exposed to radiation grow differently than unexposed seeds do?

- Does a greater amount of radiation exposure produce more mutations?

Equipment

flowerpots (5)

packet of nonirradiated seeds

packets of irradiated seeds (4)

potting soil

LAB 5D OBJECTIVES

- Compare the growth patterns of irradiated and nonirradiated seeds.

- Relate the rate of mutation to the amount of radiation.

FOR THE LONG HAUL 

This lab activity works best as a group project. It may take up to two weeks for noticeable differences to occur among the groups. The seeds used in this investigation usually take 4–12 days to germinate. Consider planting the seeds as part of one lab period and then making observations for small portions of two other lab periods.

IRRADIATED SEEDS 

Irradiated seed kits are available from science supply companies. Many come with five packets of seeds: one control packet and four packets of seeds exposed to various levels of radiation.

MUTANT PLANTS **61**

NOTES

1. Will seeds exposed to high amounts of radiation grow better or worse? State your answer in the form of a hypothesis.

   Answers will vary.

2. Describe any unusual seedlings.

   Answers will vary.

3. Did exposure to radiation increase or decrease how tall the seedlings grew?

   Answers will vary (probably decrease).

4. Did exposure to radiation increase, decrease, or have no effect on the number of unusual seedlings?

   Answers will vary (probably increase).

5. Did more or fewer seeds germinate with increased radiation?

   Answers will vary (probably fewer).

6. Why was it important to include a control group in your observations?

   The control group (the nonirradiated seeds) was needed to compare with the groups that were treated with radiation.

7. Why was it necessary to place all the flowerpots in the same conditions (temperature and light)?

   Controlling conditions ensured that observed differences in the plants were not due to differences in growing conditions, such as temperature and light.

8. How do you think radiation caused the differences that you have observed?

The radiation may have caused mutations in the seeds' genes or may have destroyed cells that are needed to produce certain chemicals or structures in the seedlings.

9. Was your hypothesis correct?

Answers will vary.

10. Do you think that giving the irradiated seeds a better environment, such as richer soil or fertilizer, would help them grow as well as the control group? Explain.

Answers will vary. Depending on the degree of damage caused by the radiation, it is unlikely that a better environment would cause the irradiated seeds to grow better than the control group. It could, though, make them grow better than irradiated seeds not placed in an enriched environment, at least temporarily.

Going Further

11. On the basis of your observations, do you think that mutations increase a plant's ability to grow and survive? Explain.

Answers will vary. Most likely they do not.

12. If irradiated plants were planted in the wild, how would natural selection affect them?

Answers will vary. Natural selection would likely result in the mutated plants' dying out.

13. On the basis of your observations, evaluate the idea that mutations produce new traits, resulting in one type of organism evolving into another.

Answers will vary. Students should point out that mutations appear to injure organisms; so rather than helping one type of organism to evolve, the mutations would likely kill the organisms instead.

14. Describe an experiment that would determine whether the offspring of the irradiated seeds would show the same unusual characteristics.

Answers will vary. Student should suggest observing subsequent offspring generations of the irradiated seeds. This would help differentiate between germ mutations and somatic mutations. Germ mutations could persist in subsequent generations, but somatic mutations and characteristics caused by the environment (and not mutations) would not persist in the offspring (unless they were exposed to those same environmental conditions).

Table 1

Group	Number of seeds planted	Number of seedlings alive after ____ week(s)	Average seedling height (cm)	Number of unusual seedlings
#1 (Control)	12	10	3.1	0
#2 (______)	12	9	0.2	5
#3 (______)	12	4	0.0	4
#4 (______)	12	4	0.0	4
#5 (______)	12	0	0.0	0

SAMPLE DATA 

The data in Table 1 is sample data. Your students' data may be different.

ODD SAMPLE DATA

The sample data in Table 1 looks very odd. The mutations in Pots 2–4 allowed some of the seeds to germinate but the seedlings never grew above the level of the dirt. As a result, although they continued to live, their height was 0 cm above the ground. Be aware that this may happen to your students' seedlings.

LAB 5E

MAKING A NEW GENE

Modeling Genetic Mutations

According to the evolutionary model, mutations produce new and different genes that result in new traits. According to this model, a robin is different from a leopard gecko because its ancestors inherited and experienced different mutations.

One way to model what it would take to produce a new gene by random mutations is to compare a gene to a sentence. The sentence used in this investigation (in Table 1) consists of only three-letter words . Recall from Chapter 3 that genes are made of nucleotides that are read in groups of three called *codons*.

Procedure

1. What does the sentence represent in our model?

The sentence represents a segment of DNA that codes for a

particular protein.

❶ Use a random number generator to randomly select which letter position (numbered 1–12) in the sentence will mutate. Record this number in the *Position to Change* column in Table 1.

2. Why are you using a random number generator to simulate the work of mutations?

Mutations in the real world are random.

❷ Use a random letter generator to randomly select a letter to replace the letter that will mutate. Record this letter in the *New Letter* column of Table 1.

3. What do the letters in the sentence in the table's header represent in our model?

The letters represent nucleotides in a segment of DNA.

❸ Rewrite the sentence in the first row of Table 1, substituting the randomly selected letter for the letter in the position that you selected with the random number generator.

❹ Beginning with the newly formed sentence from the previous turn, repeat Steps ❶ – ❸ another ten times.

NAME:

DATE:

Key Questions 

- How can changes to a sentence model changes to DNA?

- How do random changes affect the meaningfulness of a sentence?

Equipment E

random number generator

random letter generator

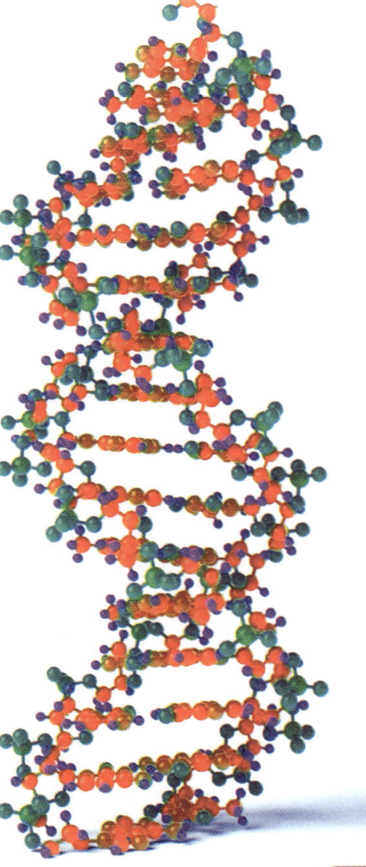

LAB 5E OBJECTIVES

- Use changes to a sentence to model changes to DNA.

- Examine the effect of random changes on an information-rich system.

RANDOM GENERATORS 

Random number and letter generators are readily available online and as apps. It's possible that the random letter generator may return the same letter to replace the letter that will mutate. This will not hurt the accuracy of this lab activity.

NOTES

4. Does the final sentence communicate anything meaningful?

Answers will vary. It probably does not.

5. What does this new sentence represent in our model?

This new sentence represents a segment of mutated DNA that no longer codes correctly.

SAMPLE DATA

The data in Table 1 is sample data. Your students' data may be different.

Table 1

Position to Change	New Letter	1 H	2 I	3 S	4 D	5 O	6 G	7 W	8 A	9 S	10 B	11 I	12 G
12	r	H	i	s	d	o	g	w	a	s	b	i	r
6	i	H	i	s	d	o	i	w	a	s	b	i	r
9	z	H	i	s	d	o	i	w	a	z	b	i	r
1	x	x	i	s	d	o	i	w	a	z	b	i	r
12	j	x	i	s	d	o	i	w	a	z	b	i	j
11	w	x	i	s	d	o	i	w	a	z	b	w	j
6	x	x	i	s	d	o	x	w	a	z	b	w	J
7	e	x	i	s	d	o	x	e	a	z	b	w	J
9	v	x	i	s	d	o	x	e	a	v	b	w	J
10	s	x	i	s	d	o	x	e	a	v	s	w	J
7	z	x	i	s	d	o	x	z	a	v	s	w	J

6. Did some letter positions never mutate?

Answers will vary. Usually some letter positions do not mutate.

7. Did some letter positions mutate more than once?

Answers will vary. Some letter positions may mutate more than once.

8. How many mutations (random letter substitutions) would it take for you to make a new, meaningful sentence? Explain.

Answers will vary. It is almost impossible for a new intelligible

sentence to form through completely random changes.

9. If the sentence used in this investigation were twice as long, would it be *more likely* or *less likely* that a new correct sentence could be produced by this type of mutation?

less likely

10. Genes consist of thousands of nucleotides (letters). How do you think this would affect the likelihood of a mutation making a new, beneficial gene?

It would be very unlikely, even impossible.

LIMITS OF THIS MODEL 

Each letter (*H*, *I*, *S*, etc.) represents a nucleotide. Three nucleotides make a codon (a word—*his*, *dog*, *was*, *big*).

One limitation of this model is that the four nucleotides found in DNA are represented by twenty-six letters in the model. This means that there are many more alphabetic codons possible in the model than the sixty-four possible with four nucleotides. Additionally, different codons sometimes code for the same amino acid. Even so, the model is valid for illustrating that the mutation of one gene into a completely different, beneficial gene is highly unlikely. The limits on making a correct sentence are balanced by using a sentence only four codons long. Genes are hundreds of codons long.

There are some initial substitutions that will work to make correct sentences. The likelihood of this randomly occurring is very low.

REVIEW 6A

MICROBES

NAME:

DATE:

Most prokaryotes are bacteria, but some are archaea. Archaea and bacteria are both microscopic, unicellular organisms. There are, however, distinct differences between them. In fact, they are classified in different domains.

Fill in Table 1, comparing these two types of organisms on the basis of the information on page 122 of your textbook. Focus on their cell walls, their genes, and whether they have a nucleus.

Table 1

Archaea and Bacteria

	Archaea	Bacteria
Cell walls	no peptidoglycan	peptidoglycan
Genes	similar to eukaryotes	different from eukaryotes
Nucleus	absent	absent

MICROBES **69**

REVIEW 6A OBJECTIVE

- Contrast the kingdoms Archaea and Bacteria.

NOTES

REVIEW 6B

VIRUSES

NAME:

DATE:

Viruses are tricky. Scientists argue about whether they are alive. To help you understand them, fill out the concept definition map below with information from pages 127–28 of your textbook. If you need help preparing a concept definition map, see Appendix B of your textbook (p. 445).

WHAT IS IT?

It is a microscopic structure with a nucleic acid and a protein coat.

DESCRIBE IT.

It has a capsid.

It is a nucleic acid.

It uses other cells to reproduce.

VIRUS

WHAT ARE SOME EXAMPLES?

flu virus

cold virus

WHAT IS IT *NOT*?

cell

bacterium

archaea

eukaryote

REVIEW 6B OBJECTIVES

- Label the structures of a virus.

- Explain why viruses are classified differently than living organisms are.

ANSWERS WILL VARY 

Students may come up with different statements to fill in the boxes of the concept definition map. Accept them if they are reasonable.

NOTES

LAB 6C

THE GLOB THAT ATE THE WORLD

Graphing Bacterial Growth

HOW QUICKLY DOES A COLONY OF BACTERIA GROW?

As you learned in your textbook, some common bacteria species reproduce by binary fission every twenty minutes. If bacteria didn't die, they could very quickly take over the world! In this lab activity, you will calculate and graph a population of bacteria starting with just one cell. Assume that the bacteria divide without any limitations and that no bacteria die during the process.

Procedure

1. Do some research on what resources a bacterial population needs. What are some factors that limit the size of a bacterial population?

   availability of food, availability of oxygen, excessive wastes,

   temperatures too high or too low, lack of moisture, and a pH too

   high or too low

❶ Calculate how many bacteria could be produced after 5 hours by filling in the total at each interval in Table 1. Some totals have been provided to help you check your work.

❷ Copy the data into an Excel spreadsheet and graph the data using a scatterplot. Add an exponential trend line.

Use your graph to answer Questions 2–3 below.

2. Estimate the amount of time that it took the population to reach 12,000 bacteria.

   approximately 4:30

3. Calculate how long it would take the bacteria population to reach one million.

   about 6:39 (Accept any answer

   between 6:35 and 6:45.)

NAME:

DATE:

Key Questions

* How fast do bacteria reproduce?
* Is it possible to estimate a population of bacteria?

Equipment

calculator

computer

Excel® spreadsheet

LAB 6C OBJECTIVES

* Collect data on bacterial growth rate.
* Plot data on bacterial growth on a graph.
* Analyze a completed graph to draw conclusions about bacterial growth.

GRAPHING 

The lab activity instructs students to graph the bacteria population using graphing software like Microsoft Excel®. Some may need help graphing data points and adding a trend line.

If your students do not have access to this type of software, they can graph and do the calculations by hand. But graphing an exponential function by hand is much more difficult, so consider that when grading.

AXIS LINES 

Some students may need help adjusting axis lines in order to answer Questions 2 and 3.

Under the x-axis settings, fix the minimum bound to 0.0, the maximum bound to 0.3, and the major unit to 0.02088. These settings should provide an axis with the major units on every half hour. You may have to adjust this slightly.

NOTES

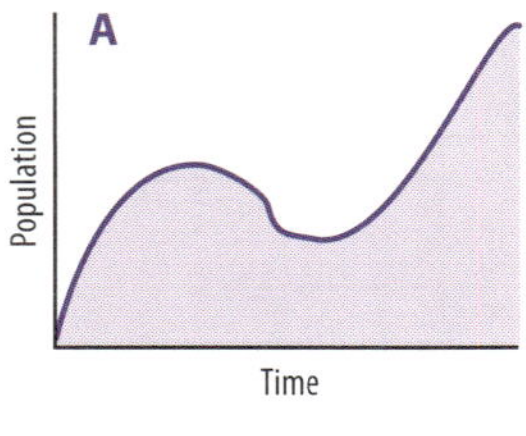

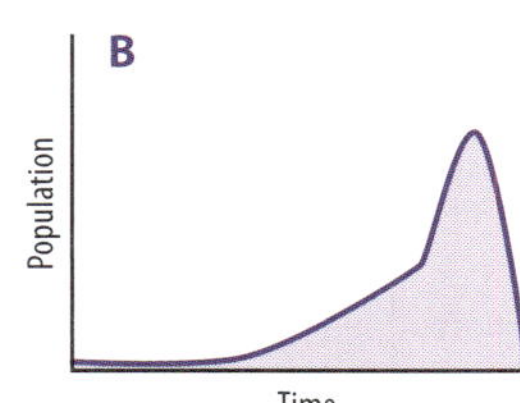

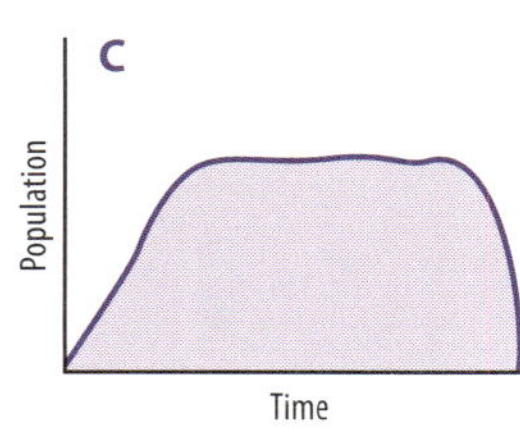

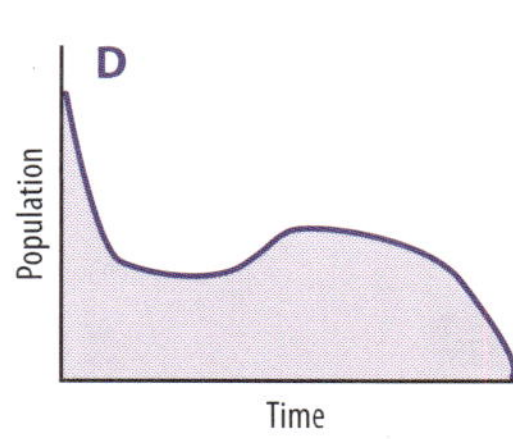

3 Examine the four graphs at left. Read the descriptions that follow and match each one with the letter of the graph that best shows what happened to the bacterial population.

___B___ **4.** The starting population grew slowly until extra food was provided. Then it grew rapidly for a short time until it ran out of oxygen and died.

___A___ **5.** The starting population grew rapidly and then began to die due to lack of oxygen. The microbiologist opened the lid of the container to allow more air to get into the colony. The colony began to grow again.

___C___ **6.** The initial population grew rapidly and then reached a point at which cells were dying as fast as they were dividing. After continuing in this balanced state for a time, excess wastes caused the whole colony to die.

7. One of the graphs was not described. What scenarios can you think of that might account for the results seen in this graph?

Answers will vary. Possibly an established bacterial colony had run out of food and the bacteria were dying. Then the scientist added a small amount of food, causing them to grow again for a short time. When that food supply was used up, the colony died. Students may come up with other scenarios as well.

8. Sketch the shape of a graph showing the population of bacteria growing on a piece of rotting fruit. Make sure that you label your axes. Begin the graph when the first bacteria cell landed on the fruit and end the graph when the fruit has completely rotted away.

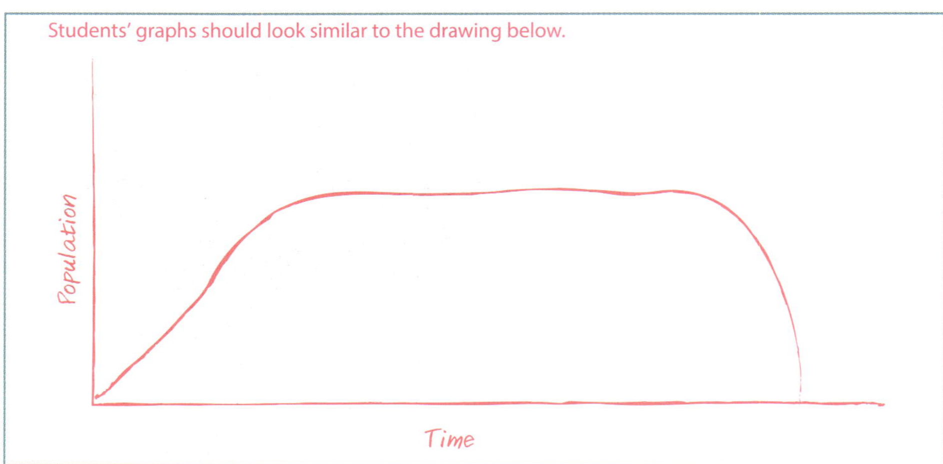
Students' graphs should look similar to the drawing below.

SKETCHING GRAPH SHAPES ✓

Make sure students realize that Question 8 is not asking them to plot a graph from data. They are to sketch a basic shape, so their sketch will probably be somewhat different from the answer shown. A student's answer is correct if it shows the same basic shape.

9. Why is it important for a population of bacteria to be able to increase rapidly?

Increasing rapidly allows bacteria to break down dead matter quickly.

Table 1

Bacterial Growth

Time (hr:min)	Number of Bacteria
0:00	1
0:20	2
0:40	4
1:00	8
1:20	16
1:40	32
2:00	64
2:20	128
2:40	256
3:00	512
3:20	1024
3:40	2048
4:00	4096
4:20	8192
4:40	16,384
5:00	32,768

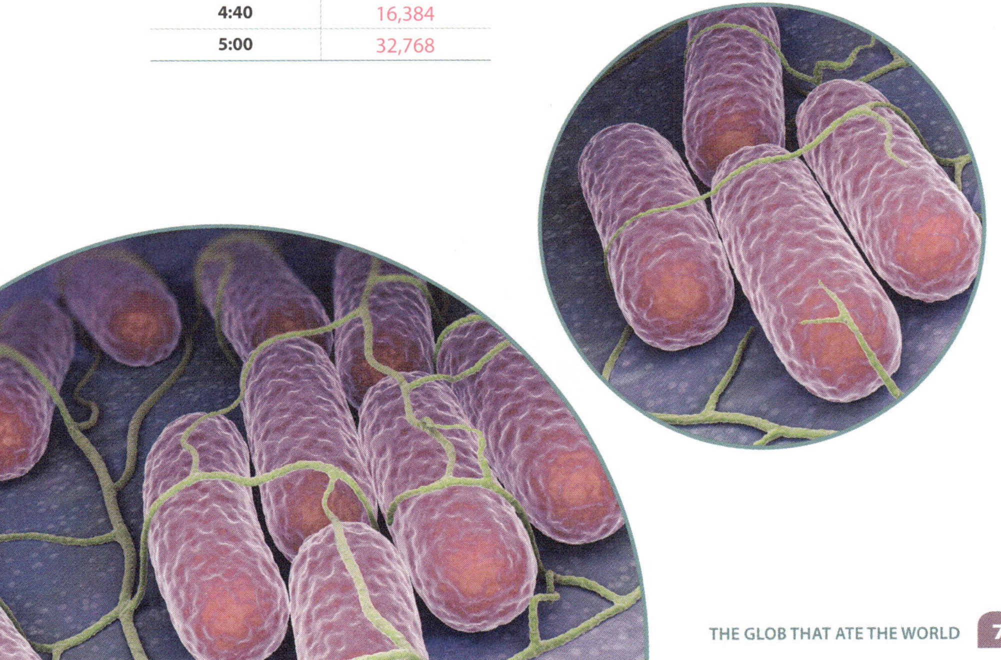

LAB 6D

HERD IMMUNITY

Modeling the Effects of Vaccines on the Spread of Disease

CAN VACCINATION PROTECT PEOPLE WHO HAVE NOT BEEN VACCINATED?

Diseases can sweep through a family, a town, or even the world. Some diseases such as influenza continue to appear every year with predictable regularity. Other diseases, such as smallpox, polio, and measles, have become much less common through the use of vaccines.

One benefit of vaccinations is herd immunity. If enough people in a population are vaccinated, unvaccinated individuals will also be protected. This is good news because in any population some people cannot be vaccinated due to health issues.

In this activity, you will model the spread of disease and the effect of vaccines on the spread of the disease. You will be using test tubes to represent people. Let's explore how disease spreads using this model.

Procedure

SETUP

1. Arrange 3 test tube holders with 10 test tubes in each. Label the groups Group 1, Group 2, and Group 3.

2. In each of the three groups, number the test tubes 1–10.

3. In each of the three groups, fill Tube 1 with 10 mL of 2% starch solution.

4. In Group 1, fill the other nine test tubes with 10 mL of water.

5. For Group 2, use a random number generator to generate four random numbers between 2 and 10, inclusive. Fill the test tubes labeled with those four numbers with 5 mL of water and 5 mL of 2% amylase solution each.

6. In Group 3, generate eight random numbers between 2 and 10, inclusive. Fill the test tubes labeled with those eight numbers with 5 mL of water and 5 mL of 2% amylase solution each.

The enzyme amylase breaks down starch into glucose.

NAME:

DATE:

Key Questions 

- How do diseases spread through a population?

- How does vaccination affect the spread of disease in a population?

Equipment

starch solution (2%) (30 mL)
amylase solution (2%) (48 mL)
tincture of iodine
test tubes (30)
test tube holders (3)
graduated cylinder, 10 mL
disposable pipettes (31)
37 °C water
hot plate
random number generator
goggles

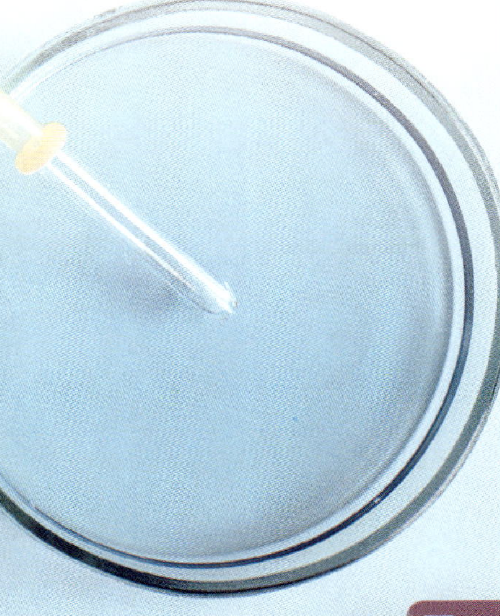

LAB 6D OBJECTIVES

- Model the spread of disease in a population.

- Model the effect of vaccination on the spread of disease in a population.

LAB PREP 

This lab activity works well for multiple groups. Give each group a population with a different vaccination rate and have them pool their data.

STARCH SOLUTIONS 

Starch solution is easily purchased from science supply companies. Amylase is typically sold as a soluble powder. Follow the instructions on the container for dissolving the powder into solution.

HERD IMMUNITY **77**

NOTES

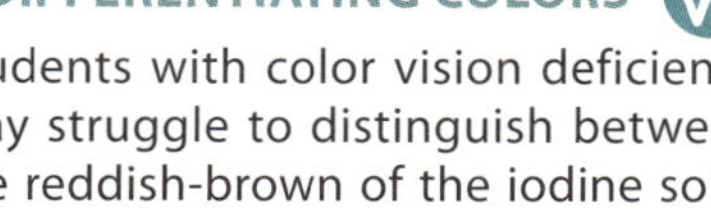

DIFFERENTIATING COLORS ✓

Students with color vision deficiency may struggle to distinguish between the reddish-brown of the iodine solution and the purple to black color of the iodine in the presence of starch. If one of your students has color vision deficiency, be sure that he is paired with a student with normal color vision.

Since both the amylase solution and the starch solution are cloudy, they can distort the color of the iodine. It might be helpful to have examples of iodine in water, iodine in starch solution, and iodine in amylase solution for students to compare their results against.

SAMPLE DATA ✓

The data in Table 1 is sample data. Your students' answers may be different.

Iodine dyes starch black, blue, or purple, depending on the concentration of the starch. If starch is not present, iodine remains a reddish orange color.

Table 1
Infected Tubes

Group	Number of Infected Tubes
1	10
2	5
3	1

7 Fill each of the remaining test tubes in Groups 2 and 3 with 10 mL of water.

8 Place a pipette in each test tube.

1. What does each tube of starch solution represent in this model?

a person who has been infected with a disease

2. What does the amylase solution represent in this model?

a vaccine

3. Which population will have the lowest infection rate? Which population will have the highest infection rate? Write your answer in the form of a hypothesis.

Answers will vary.

INFECTION!

1 For each group, transfer 2 mL of the solution in Tube 1 to each of Tubes 2, 5, and 8.

2 For each group, transfer 2 mL of the solution from Tube 2 to Tube 3. Then transfer 2 mL of the solution from Tube 3 to Tube 4.

3 For each group, transfer 2 mL of the solution from Tube 5 to Tube 6. Then transfer 2 mL of the solution from Tube 6 to Tube 7.

4 For each group, transfer 2 mL of the solution from Tube 8 to Tube 9. Then transfer 2 mL of the solution from Tube 9 to Tube 10.

5 Allow the test tubes to sit for 2 minutes.

4. After the transfers, do you think that you will find starch in Tube 3 in Group 1? Explain.

Yes. The starch solution is merely diluted in Tube 2, so starch is still present. That starch solution will be passed to Tube 3.

5. If Tube 2 in Group 3 originally contained amylase, do you think that you will find starch in Tube 3 after the transfers? Explain.

No. The amylase in Tube 2 will break down the starch, so only glucose will be passed to Tube 3.

6 Add 5 drops of iodine to each test tube. If starch is present, the iodine will dye it blue or purple.

7 In Table 1, indicate the number of test tubes in each group "infected" with starch. Include Tube 1 (the tube with starch solution).

6. What does each transfer from one tube to another represent in this model?

a pathogen being passed from one person to another

7. Which group had the least number of infected tubes?

See TE margin for answer.

8. Was your hypothesis correct?

Answers will vary.

Going Further

9. If each of the groups above represents a city of 500,000 people, how many people in each city would contract the disease?

See TE margin for answer.

10. In countries with good health care, the average death rate from a measles outbreak is around 0.2%. Use the infection numbers that you calculated in Question 9 to determine how many people would likely die in each population.

See TE margin for answer.

11. How does vaccinating a large number of individuals in a population help protect unvaccinated individuals?

Answers will vary. Individuals who have been vaccinated will not contract or spread a disease. If a disease is less likely to be spread, unvaccinated individuals are less likely to contract the disease.

12. Explain how getting vaccinated can fulfill Jesus' command to love your neighbor toward those who cannot be vaccinated.

Answers will vary. Getting vaccinated can prevent the spread of disease, helping protect those who cannot be vaccinated.

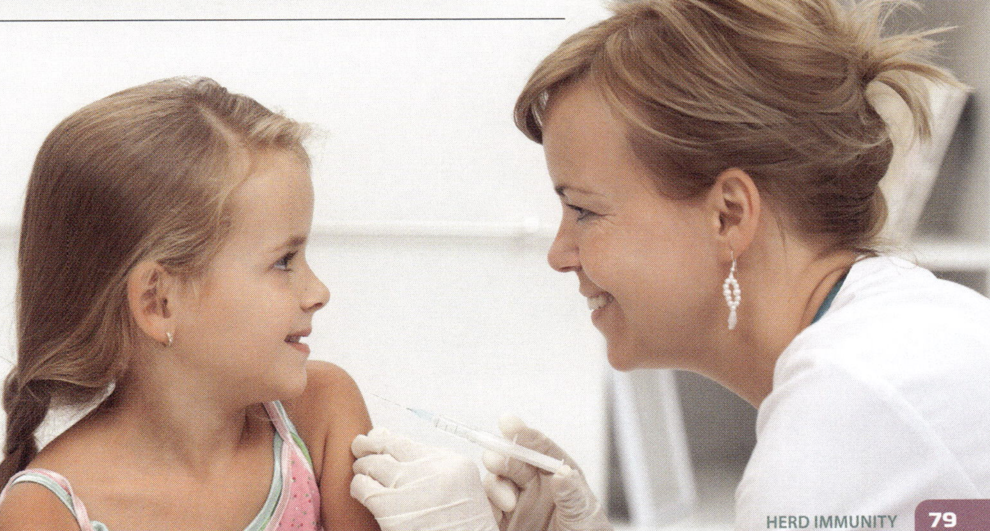

Question 7 Answer

Answers will vary. In most cases, Group 3 will have the least number of infected tubes. In a few cases, Group 2 and Group 3 will have an equal number of infected tubes.

Question 9 Answer

Answers will vary. The number of infected tubes in each group should be divided by 10 to find an infection rate and then multiplied by 500,000.

Question 10 Answer

Answers will vary. The number of infected persons calculated in Question 9 should be multiplied by 0.002.

VACCINATION 

Receiving vaccines in any population requires a personal choice and action. We have to understand that we never make a decision in a vacuum and our personal choices have an impact on those around us.

HERD IMMUNITY 79

REVIEW 7A

CLASSIFYING PROTISTS

NAME:

DATE:

Green pond scum, dangerous *Giardia*, weird slime molds—protists are a diverse group! But they can be roughly divided into groups according to what other organisms they resemble.

For each protist pictured below, indicate whether it is animal-like (*A*), plantlike (*P*), or fungus-like (*F*). Then, for each image, indicate what characteristics you used to classify the organism. Indicate any characteristics that might cause someone to classify it differently.

1. Note the color of these algae that live in water.

   P; Most students will probably indicate that the fact that algae are green, make their own food, and do not absorb or capture food led them to classify them as plantlike. No major features give reason to place these algae in another category.

2. Note the cilia on the exterior of this *Paramecium* that allow it to move through the water to capture prey.

   A; Most students will probably indicate that they classified the *Paramecium* as animal-like because it has independent movement and cannot make its own food. No major features give reason to place *Paramecia* in another category.

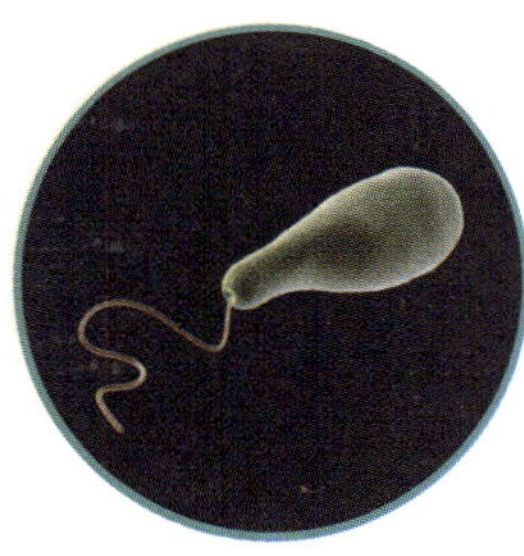

3. Notice the flagellum on this *Euglena* that allows it to move through the water. It also has chloroplasts.

   A; Answers will vary. Students will probably mention that they classified the *Euglena* as animal-like because of the flagellum, which indicates the ability to move. Some students may mention that *Euglena* could be classified as plantlike because of their chloroplasts.

REVIEW 7A OBJECTIVE

- Classify protists on the basis of major characteristics.

ALTERNATIVE ANSWERS 

If a student has a different answer to one of the questions, give him credit if he can defend it logically or scientifically.

NOTES

4. *Physarum polycephalum* lives on dead vegetation.

F; Answers will vary. Students will probably mention that the fungus-like appearance of this slime mold led them to classify it as fungus-like. Students may mention characteristics such as the ability of some slime molds to move that might give reason to classify them as animal-like.

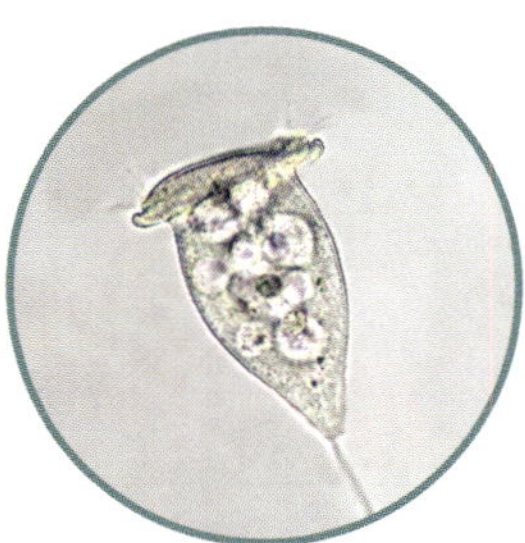

5. *Vorticella* lives in water, is attached by a stalk to a substrate, and uses cilia to capture its food.

A; *Vorticella* is one of the more difficult organisms to classify in this exercise. Most students will probably classify it as animal-like. Some students might classify *Vorticella* as fungus-like since it is sessile.

6. Kelp lives in shallow water, where it can absorb sunlight for photosynthesis.

P; Most students will probably indicate that the fact that kelp are green and make their own food led them to classify them as plantlike. No major features give reason to place kelp in another category.

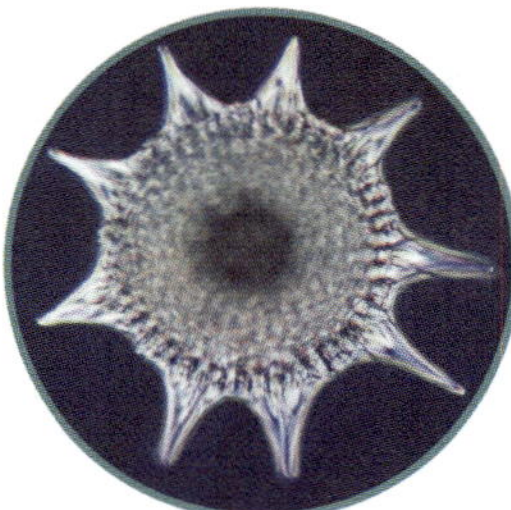

7. Radiolarians have skeletons made of minerals but also have pseudopods that they use to capture food.

A; Most students will probably classify radiolarians as animal-like because of their pseudopods and method of capturing food. No major features give reason to place radiolarians in another category.

REVIEW 7B

MUSHROOM STRUCTURE

NAME:

DATE:

- Identify the major structures of fungi.

The harmless false death cap shown below looks very similar to the deadly poisonous death cap (see page 150 in your textbook). For that reason, people don't eat it. There's not a good way to tell the difference between it and the real death cap, so it's just not worth the risk.

In Questions 1–5, match each description below with the letter of the part of the fungus being described. Then write the name of the structure.

1. structures designed for reproduction

   c; spores

2. long fungal filaments

   a; hyphae

3. the structure that supports the gills

   b; cap

4. the structures that produce reproductive structures

   d; gills

5. the stalk of the mushroom

   e; stipe

6. Is the false death cap an imperfect fungi? Explain.

   No. It reproduces using spores.

7. Round fungi called puffballs lack gills. But their spores are produced by structures with the same shape as those that produce spores on the gills of mushrooms. Which group of fungi do puffballs belong to?

   club fungi

8. Argue against this organism being classified as an animal or plant.

   It cannot be an animal because it has cell walls. It cannot be a

   plant because its cell walls contain chitin, not cellulose.

NOTES

LAB 7C

AMOEBAS IN ACTION

Observing a Protist

WHAT IS AN AMOEBA LIKE?

Most protists are microscopic. You can't usually see them unless there are thousands of them all together in a group, like pond algae. But with a microscope, you can see an individual protist. There's a whole world that is invisible to you!

Sometimes it's a friendly world, and sometimes it's not. For example, the amoeba *Naegleria fowleri* is called the "brain-eating amoeba." It normally lives peacefully in warm lakes where it eats bacteria. But on extremely rare occasions, it finds its way up a swimmer's nose and causes a deadly brain infection.

In this lab activity, you will use a microscope to observe a live but harmless amoeba. Watch out for amoebas in action!

Procedure

1. Use a pipette to obtain a tiny sample of the amoeba culture. Don't worry—this one won't eat your brain!

2. Place the sample in the depression of your microscope slide. Place a cover slip on the slide.

3. Observe your amoeba culture under the microscope. Use low power to locate an amoeba. After you have located it, you may want to observe it under a higher power to see more detail.

4. Locate the nucleus and any vacuoles. If any of these vacuoles are changing in size, they are most likely contractile vacuoles. (Contractile vacuoles fill and empty their contents regularly.)

1. What do you think the contractile vacuoles are doing? (*Hint*: Osmosis causes water to flow into the amoeba.)

   Answers will vary. Accept any reasonable answers. The contractile vacuoles push water out of the amoeba to prevent the amoeba from bursting.

NAME:

DATE:

Key Questions
- What does an amoeba look like?
- What structures and functions of an amoeba can I observe?
- What kind of protist is an amoeba?

Equipment
culture of living amoebas
pipettes (several)
blank concavity microscope slides (several)
cover slips (several)
microscope

LAB 7C OBJECTIVES

- Describe a living amoeba.
- Describe the structure and function of a typical protist.
- Classify an amoeba on the basis of its structure and function.

OTHER PROTISTS 

You may substitute another protist for the amoeba. But amoebas are some of the easiest protists to observe since they move slowly. Other protozoans often move too quickly to be seen easily. Algae don't move at all.

You might consider purchasing several other protists in addition to amoebas for your students to view through their microscopes. Some possibilities include *Euglena*, *Paramecium*, *Spirogyra*, and *Volvox*. You might also consider obtaining a sample of pond water and having students examine the organisms within it under a microscope.

KEEP IT FRESH! 

Most science supply companies will ship living specimens so that they arrive on a particular date. Protozoan and algal cultures need to be fresh.

TEACHER DEMONSTRATION 

Making live mounts can be a very challenging skill. You could do this lab activity as a teacher demonstration, if you want to focus on the content more than the laboratory skills.

85

NOTES

You may allow your students to post pictures or even take a video of an amoeba. But the practice of sketching an amoeba helps students observe more closely. Direct your students to Appendix C in their Student Activities for instructions on making biological drawings.

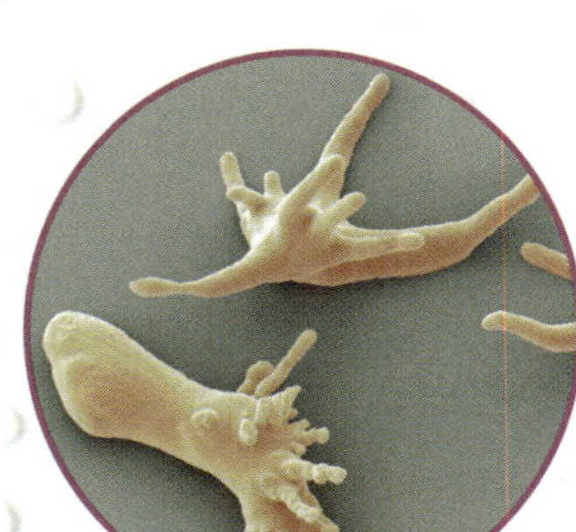

This scanning electron micrograph of an amoeba shows its external shape.

⑤ Draw the basic shape of your amoeba in the first box below. Wait two minutes and then draw the basic shape of your amoeba again in the second box.

| | |
| | |

2. Which of the following terms apply to amoebas: eukaryotic, multicellular, flagella, mitosis, pseudopods, unicellular, algae, prokaryotic, fungi?

eukaryotic, mitosis, pseudopods, unicellular

3. If you saw your amoeba moving, describe its movement.

Answers will vary. If they saw movement, students should describe a flowing of cytoplasm to form long, thin pseudopods. It may be described as a very slow "crawling" movement.

4. How is the movement of the amoeba different from the movement of other organisms in the same kingdom?

Amoebas move by pseudopods. Other protists, such as *Paramecia*, move by means of cilia. Still others, such as *Giardia lamblia*, have flagella for movement.

5. Would you classify the amoeba you observed as plantlike, animallike, or fungus-like? Explain.

Because of their ability to move and their predatory habits, amoebas would be classified as animal-like.

Going Further

6. Most amoebas are harmless, but some can cause disease, often fatally. How do such harmful organisms fit with God's description of His creation as "very good"?

Answers will vary. Students should mention that disease and death came as a consequence of man's sin. We can assume that these harmful species of amoebas were harmless before the Fall.

LAB 7D

THE PICKY YEAST

Temperature and Yeast

WHAT IS THE OPTIMAL TEMPERATURE FOR YEAST?

Do you like donuts? How about pizza? Both of these foods rely on yeast to make the dough rise. As yeast, an imperfect fungus, digests sugar, it releases carbon dioxide. This gas causes the dough to rise.

But yeast can be a picky fungus. It works much better at some temperatures than at others. It's important for bakers to know what the best temperature is for the sugar and yeast mixture in order for the dough to rise properly. In this activity, you will investigate the effect of temperature on yeast and to see whether you can determine the best temperature range for yeast activity.

NAME:

DATE:

Key Questions

- What does yeast need to grow?
- How does temperature affect the activity of yeast?

Equipment

apple juice, 200 mL

yeast, 20 g

bottles, 20 oz (4)

stopwatch

warm water baths (2)

ice water bath

graduated cylinder, 50 mL

balloons (4)

thermometers (4)

Procedure

1. Which temperature do you think will be the best for yeast? Write your answer as a hypothesis.

   Answers will vary. Example: A temperature of 60 °C will be the

   best for yeast.

❶ Prepare an ice water bath, a warm water bath at 40 °C, and a warm water bath at 60 °C.

❷ Use the graduated cylinder to pour 50 mL of apple juice into each bottle.

❸ Place a bottle in each water bath. Leave one bottle to stand at room temperature. Place a thermometer into each bottle.

❹ Allow the apple juice time to warm or cool to the temperature of the bath.

❺ Add 5 g of yeast to each bottle and swirl the bottle, thoroughly mixing the yeast and juice.

LAB 7D OBJECTIVES

- Identify the conditions needed to nurture yeast.

- Compare the effect of temperature on the activity of yeast.

SETUP

This activity works well as a teacher demonstration or a group activity if you do not have enough water baths for all the members of your class.

You should set up the water baths before class begins.

EQUIPMENT

Sugar water is a very workable substitute for apple juice.

BURN HAZARD

A temperature of 60 °C is hot enough to burn skin, so you should supervise your students to see that they are careful around the hot water bath.

AMOUNTS

The exact amounts of apple juice and yeast are not important. But it *is* important that all the bottles contain the same amounts each of apple juice and yeast.

NOTES

Question 4 Answer

Answers will vary. Most likely the balloon on the bottle in the 40 °C warm water bath is the most inflated.

Question 5 Answer

Answers will vary. Most likely the 0 °C and the 60 °C bottles will have uninflated balloons.

Question 6 Answer

Answers will vary. The student should indicate the temperature that resulted in the most inflated balloon. Most likely it will be 40 °C.

2. Why are you mixing apple juice with the yeast?

Yeast feeds on sugar.

6 Place a balloon over the mouth of each bottle. Using balloons of different colors might help distinguish the different bottles.

3. How will the balloons show the activity of yeast?

A greater amount of yeast activity will produce a greater amount of carbon dioxide, which will inflate the balloon more than less carbon dioxide will.

7 After 5 minutes, rank the balloons by how inflated they are.

4. Which balloon is the most inflated?

See TE margin for answer.

5. Which balloon is the least inflated?

See TE margin for answer.

6. On the basis of your results, which temperature do you think is closest to the optimal temperature for yeast?

See TE margin for answer.

7. Was your hypothesis correct?

Answers will vary.

8. How could this information be useful?

When using yeast in cooking, a baker could use warmers to accelerate rising. Yeast can also be activated in warm water to speed the rising process.

Going Further

9. On the basis of your results, design an experiment to find the *exact* optimal temperature for yeast.

Answers will vary. The student could suggest doing a series of experiments with temperatures near 40 °C to find the temperature that results in the most yeast activity.

MAJOR PLANT GROUPS

NAME: _______________

DATE: _______________

Use the terms from the list below to fill in the hierarchy chart showing the relationships between the groups of plants. There is only one correct way to use all the terms in this hierarchy chart. Two terms have already been placed on the chart to get you started.

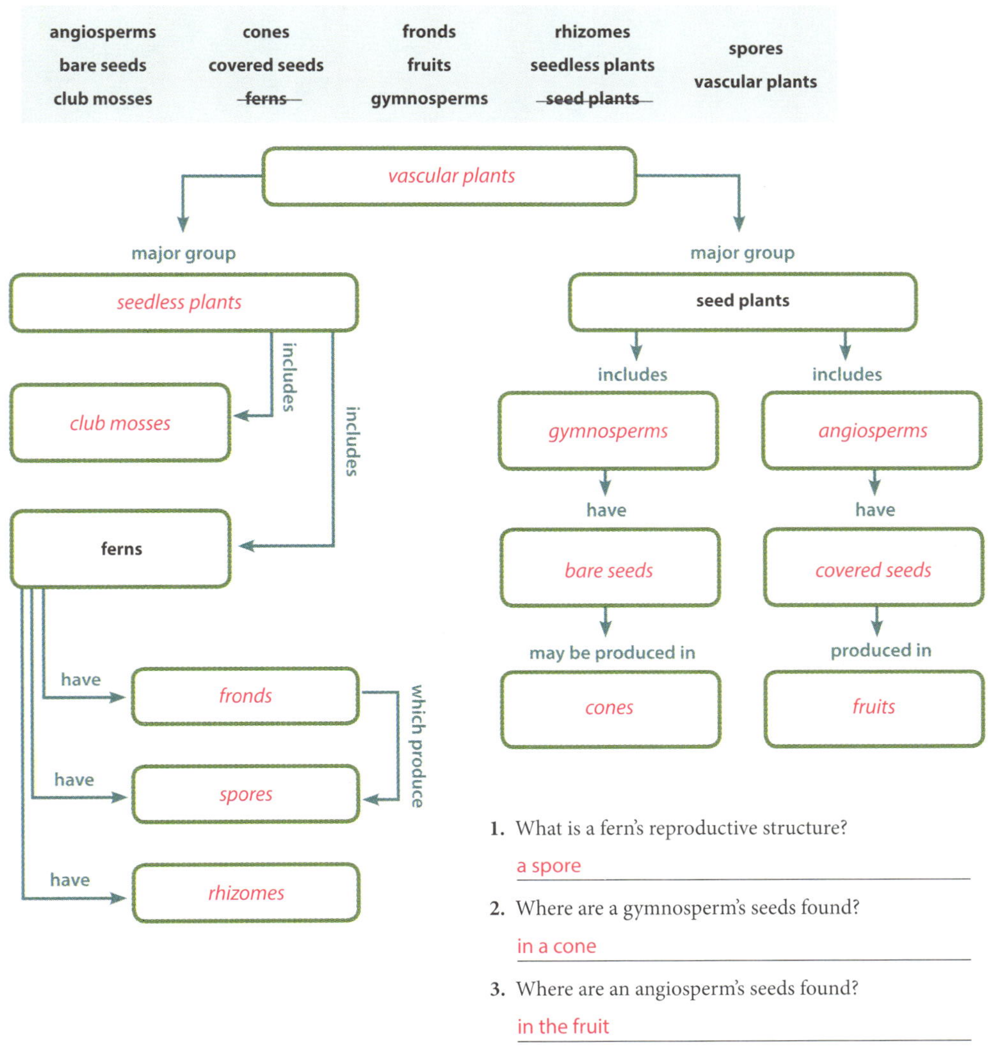

1. What is a fern's reproductive structure?
   a spore
2. Where are a gymnosperm's seeds found?
   in a cone
3. Where are an angiosperm's seeds found?
   in the fruit

REVIEW 8A OBJECTIVE

- List the main characteristics of plants.

HIERARCHY CHART 

Hierarchy charts are graphic organizers that show the relationship between concepts. They are especially useful when classifying objects or concepts. For more information on hierarchy charts, see Appendix B in the Student or Teacher Edition.

ALTERNATE ANSWER 

Some students may name horsetails instead of club mosses.

NOTES

____

____

____

____

____

PLANT ANATOMY

NAME:

DATE:

Supply the missing terms on the basis of the definitions given. Then in the diagram of the plant, label each structure that has a line pointing to it, using the number of each term as your label. One of them is done for you as an example.

REVIEW 8B OBJECTIVE

- Identify the structure and functions of roots, stems, and leaves.

1. cuticle
(thin and often shiny protective layer of a leaf)

2. annual ring
(formed by differing xylem growth rates)

3. stem
(transports substances between roots and leaves)

4. cork
(protective layer made of dead cells)

5. cork cambium
(layer of living cells that produce bark)

6. spongy layer
(tissue that contains many air spaces)

7. stoma
(an opening for the exchange of gases)

8. root
(absorbs water and nutrients)

9. vein
(vascular tissue of a leaf)

10. palisade layer
(tissue in which most photosynthesis takes place)

NOTES

____

____

____

____

____

____

PLANT COLLECTING GREW ON ME

Inquiring into Plant Characteristics

HOW CAN PLANTS BE CLASSIFIED?

Think of your local garden supply center. There you can find a section with flowers, another with shrubs, and yet another with trees. But within these general categories there are many species and varieties. There are differences between them, even though at first glance they may appear identical. By taking a closer look at plants, you may find differences that can be used to classify them.

Procedure

1. Make a collection of several different types of plants. If a plant is small, you may collect the entire plant. If the plant is a large shrub or a tree, collect only a portion of a stem.

2. Carefully observe the characteristics of the plants that you have collected.

   1. What are some of the characteristics of your plants?

      See TE margin for answer.

NAME:

DATE:

Key Question ?
- What are some of the key characteristics of plants?

Equipment E
various plant specimens

Safely Handling Plants
Be careful when handling plants. Some may cause dermatitis, like poison sumac (below), or have thorns, like blackberries. Be sure to tell your teacher if you have any significant plant allergies. In your search for plants, do not disturb any animal nests that you encounter in the field.

Poison sumac (Toxicodendron vernix)

LAB 8C OBJECTIVE

- Observe and identify plant traits that could be used to classify plants.

SOME TIPS FOR LAB 8C 

This activity will obviously require some time spent outside of class to gather specimens. You could assign this as individual projects or as a group project. Doing it as a group project may enable students to gather a broader variety of specimens. You could also do this as a class and assign each student or group a different type of plant to collect.

When collecting portions of large plants such as trees, students should collect as large a specimen as practical. A branch, for example, will have more distinguishing characteristics than a single leaf.

Alternatively, this activity can be done as a field exercise with the necessary observations made in a field notebook. Students should include a sketch or photo of each plant that they describe.

Question 1 Answer

Answers will vary depending on the types of plants that are chosen. Characteristics may include the size of the plant, the size of the leaves, the presence of flowers or fruit, the type of stem, the shape of the leaves, and the way that the leaves connect to the stem.

PLANT COLLECTING GREW ON ME **93**

NOTES

❸ Note the similarities and differences between the characteristics of your plants that you noted in Question 1. On a separate sheet of paper, compile two lists, one that describes the similarities between the various specimens and one that describes the differences.

Analysis

2. Think about how the similarities and differences that you observed in your plant specimens can be used to classify them. On the basis of what you observe, create a hierarchy chart similar to the one that you saw in Review 8A.

Answers will vary. Check to make sure that the student's chart is both functional and consistent with the similarities and differences described above.

Going Further

3. Do an Internet search using the keywords "plant classification systems." What are modern classification systems based on?

Modern plant classification systems are based on analysis of DNA and other molecules.

4. How is your plant classification system different from the modern system?

The system you created is based on the external appearances of the plants, not on differences in their DNA.

WORLDVIEW AND CLASSIFICATION 

The classification system that students create will be based on what they can easily see regarding their plants. This is how Carl Linnaeus created his classification system. Modern classification systems are based on the genetic similarities between organisms in an effort to discern presumed evolutionary relationships.

Consider having a group discussion on the merits and difficulties inherent in modern classification systems. Potential questions to consider: Is genetic similarity evidence for common ancestry? Proposed evolutionary relationships are frequently revised. What does this say about modern classification systems as tools outside the field of taxonomy?

IDENTIFYING PLANTS 

If time permits, interested students may find attempting to identify their plant specimens to be rewarding. Have suitable field guides available and be prepared to coach your students on how to use them.

LAB 8D

CRUNCH TIME

Exploring Turgor Pressure

WILL SALT WATER MAKE POTATO STRIPS CRISPER?

A nifty kitchen trick for making vegetable sticks or slices nice and crunchy for a party tray is to soak the vegetables in ice-cold water. But will salt water do the job too?

NAME:

DATE:

Key Question 

- Is turgor pressure affected by salt water?

Equipment

shallow dishes or bowls (6)

tap water

salt water

knife

cutting board

balance

fresh potato

LAB 8D OBJECTIVE

- Analyze the effects of fresh and salt water on turgor pressure.

NOTES ON MATERIALS

For best results, your saltwater solution should be saturated. Add salt to your solution until no more will dissolve (about 35 g per 100 mL at room temperature).

The potatoes must be fresh. Frozen or old potatoes will not work.

To save class time, you may prepare the potato strips in advance and store them in cold water. Drain the water from the potato strips prior to beginning the activity.

1. Write your hypothesis about the effect of a saltwater soak on vegetable crispness.

   Answers will vary. Example: Vegetables soaked in salt water will

   be less crispy than those soaked in tap water.

Procedure

❶ Pour tap water into three dishes to a depth of about 5 cm. Label these dishes *Tap Water* and number them 1–3.

❷ Place the same amount of salt water into the remaining three dishes. Label these dishes *Salt Water* and number them 4–6.

❸ Cut six thin strips of potato about 5 mm thick and at least 5 cm long. You should be able to completely submerge them in the water in the dishes. The potato strips should be as nearly identical as you can make them.

2. Why is it important that the potato strips be as similar as possible?

   This will make it easier to compare the effect of the saltwater

   soak on turgor pressure.

NOTES

If your salt solution is saturated or near saturated, the potato strips will likely float in the salt water rather than sink to the bottom of the dish as the potato strips in the tap water will. This is a result of the higher density of salt water and will not affect the results of the experiment.

4 Measure the mass of one of the potato strips and record the mass in the first row of the *Mass Before Soaking* column of Table 1.

5 Place the potato strip in Dish 1. Make sure that it is completely covered; add more tap water if necessary.

3. Why is it important to make sure that the potato strip is completely covered?

Complete coverage maximizes the surface area of the potato strip that is in contact with the water.

6 Repeat Steps **4** and **5** for the remaining potato strips, recording your data in the appropriate rows. Place one strip each in Dishes 2–6.

7 Leave the strips undisturbed for 30 minutes.

4. Why is it important to do several trials for both tap and salt water?

Multiple trials help us to be more certain about our results by seeing whether they are repeatable.

8 After 30 minutes, take the potato strips out of the dishes and record your observations about the stiffness of each in the *Stiffness* column of Table 1.

9 Carefully blot away any excess water from the strips. Measure the mass of each strip and record the data in the *Mass After Soaking* column.

5. Why is it important to dry off the outside of the potato strips?

Removing excess water assures that any change in weight is due to water that the potato strips have absorbed.

CONCLUSION

6. How did the mass of each of the potato strips change?

The mass of the potato strips soaked in tap water increased slightly. The mass of the potato strips soaked in salt water decreased.

7. Did these results confirm your hypothesis? Explain.

Answers will vary. Check to make sure that each student's answer is consistent with both his original hypothesis and his observations.

8. What do you think caused the results that you described in Question 6?

The potato strips soaked in tap water *absorbed* water from their environment. The potato strips soaked in salt water *lost* water to their environment.

9. Which potato strips were stiffer, those soaked in tap water or those soaked in salt water?

The strips soaked in tap water were stiffer.

10. Use the concept of turgor pressure to explain your answer to Question 9.

Turgor pressure, which keeps plant cells rigid, is caused by water pressure inside a plant's cells. The potato strips soaked in tap water absorbed water, causing an increase in their turgor pressure. The potato strips soaked in salt water lost water, causing a decrease in turgor pressure.

Going Further

11. Take a look at the picture on the right. Use what you have learned from this activity to explain what is happening in the photo.

The tree roots absorb water, increasing the turgor pressure within their cells. The increased turgor pressure is sufficient to dislodge the concrete slabs of the sidewalk.

Table 1

	Mass Before Soaking (g)	Mass After Soaking (g)	Stiffness
Tap water Dish 1	5.72	5.90	stiff
Tap water Dish 2	5.75	5.95	stiff
Tap water Dish 3	10.85	11.08	stiff
Average of potato strips in tap water	7.44	7.64	
Salt water Dish 4	6.44	5.39	flexible, not stiff
Salt water Dish 5	5.89	4.96	flexible, not stiff
Salt water Dish 6	12.03	10.61	flexible, not stiff
Average of potato strips in salt water	8.12	6.99	

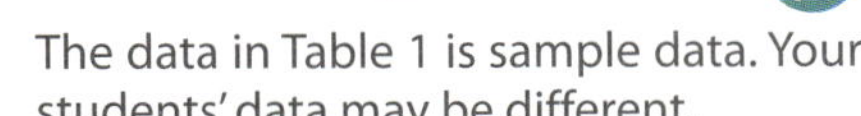

SAMPLE DATA ✓

The data in Table 1 is sample data. Your students' data may be different.

REVIEW 9A

CLASSIFYING TROPISMS

NAME:

DATE:

Have you ever seen a vine growing up a tree? How about an indoor plant growing toward a window? As you learned on page 177 of your textbook, these types of growth patterns are called *tropisms*.

Answer Questions 1–5 by indicating (1) whether the indicated image shows a positive or negative tropism (if that applies) and (2) whether the tropism displays phototropism, gravitropism, or thigmotropism. Some images may show more than one tropism.

REVIEW 9A OBJECTIVE

- Define tropism in plants.

1. What tropism(s) is the shoot above demonstrating?

positive phototropism and negative gravitropism

2. What tropism allows the vine at right to climb the pole?

thigmotropism

3. What other tropisms cause the vine to grow away from the ground?

positive phototropism and negative gravitropism

NOTES

4. The corn kernel above sprouted in the dark. What tropism causes its root to grow downward?

positive gravitropism

5. What tropism(s) are the exposed roots of this tree evidencing?

negative phototropism and positive gravitropism

6. A seed sprouts an inch beneath the soil. When does the stem begin to demonstrate phototropism?

When the stem breaks through the surface of the soil, it begins

to demonstrate phototropism. Before that time, it is guided by

negative gravitropism.

FLOWERS

NAME:

DATE:

Blooming tulips are a sign that spring has arrived. The flower of the tulip is beautiful and bright and provides the way for the plant to reproduce.

On page 185 of your textbook, you learned the parts of a typical flower. In the activity below, match each description with the letter of the structure it describes. Then write the name of the structure.

1. the structure that contains the pollen

e; anther

2. the structure that supports the anther

f; filament

3. the structure that contains the ovules

c; ovary

4. the structures that will become seeds

j; ovules

5. the often large, showy parts of flowers

h; petals

6. the female reproductive structure

d; carpel

7. the structure that contains the male gametes

k; pollen

8. the often leaflike structures that protect the developing flower

i; sepals

9. the male reproductive structure

g; stamen

10. the surface to which pollen becomes attached

a; stigma

11. the slender stalk supporting the stigma

b; style

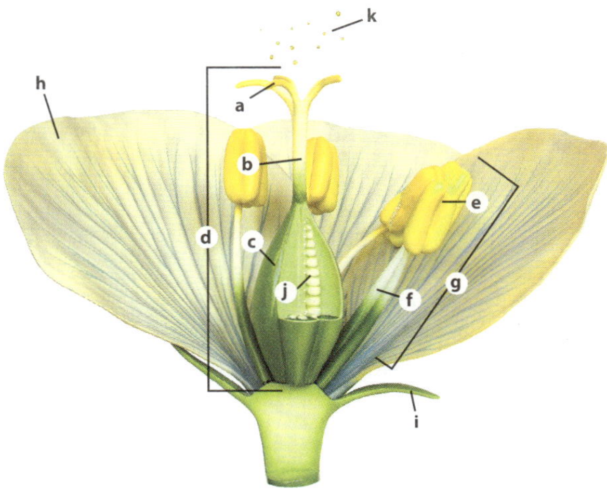

REVIEW 9B OBJECTIVE

- Compare reproduction in gymnosperms and angiosperms.

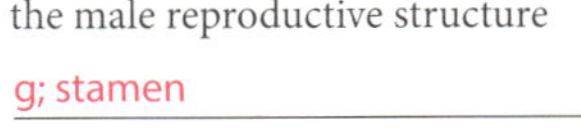

NOTES

LAB 9C

WHICH WAY IS UP?

Gravitropism in Seedlings

DOES THE WAY IN WHICH A SEED IS PLANTED AFFECT HOW ITS STEM AND ROOTS GROW?

When astronauts spend lots of time in space, they have an unexpected challenge—space sickness. It's like the motion sickness you get in a car. You see, in space, there is no up. There's no surface! Astronauts float around in the microgravity environment of the International Space Station, and they have to get used to living without gravity.

Plants are like us: they use gravity as a reference point. We all know that stems grow up and roots grow down. Gravitropism and phototropism work together to make this happen. But it can be hard to study one of these tropisms without the other.

Plant seedlings can grow for a few days without light, so it is possible to test gravitropism without phototropism confusing the results. In fact, they need this ability since they often germinate underground. If you sprout seeds in the dark, you can experiment with gravitropism.

Bean seeds have a distinct structure—it's easy to predict which side of the bean the seedling and root will grow from. Will this orientation affect the direction of growth as much as gravitropism? This is your chance to find out!

1. Do you think that the direction the seed is facing will affect the direction of growth of the stem and root? Write your answer as a hypothesis.

   Answers will vary.

   ____

Procedure

① Place several folded paper towels on a glass plate. Place rubber bands around the glass plate and paper towels to hold the towels in place.

② Place eight small, wet cotton balls on the paper towels. Then place a bean seed in the center of each cotton ball. Be sure that two bean seeds are oriented in each direction (up, right, left, or down). You should have two seeds facing up, two facing right, two facing left, and two facing down.

③ Slowly add a little water to wet the paper towels.

NAME:

DATE:

Key Question 

- Is it possible to demonstrate gravitropism?

Equipment E

paper towels (several)

glass plates (2)

rubber bands (several)

cotton balls (8)

bean seeds (8)

water

marking pencil

eyedropper

LAB 9C OBJECTIVE

- Demonstrate gravitropism in seedlings.

EQUIPMENT 

For the glass plates, use glass from picture frames or windowpanes, which are available from a hardware store or a store with custom framing services. The panes should be approximately 6 × 6 in. You may want to tape the edges.

Dry beans (e.g., lima, pinto, kidney, or navy) from a grocery store usually work well, but be sure to test them for germination before beginning this investigation.

CONCURRENT LABS ✓

Labs 9C and 9D both require several days between setup and finish. If you choose to have your students perform both lab activities, you can have them set up for both in a single lab period. Students can then make their observations for Lab 9C between the setup day and the next lab day, when they can complete both labs.

Since most of the germination-inducing methods in Lab 9D require twenty-four hours, you should have students expose their seeds to their methods the day before. Or you could have them plant their seeds the next day.

Weekends will likely prevent students from inspecting their seed tray on each day. If the seeds are planted on a Friday, it is unlikely that any seeds will germinate before the following Monday.

NOTES

____

____

____

____

____

2. What is the purpose of the water?

The purpose of the water is to cause the seeds to germinate.

3. What is your independent variable?

the placement of seeds in planting

4. What is your dependent variable?

the direction of root and shoot growth

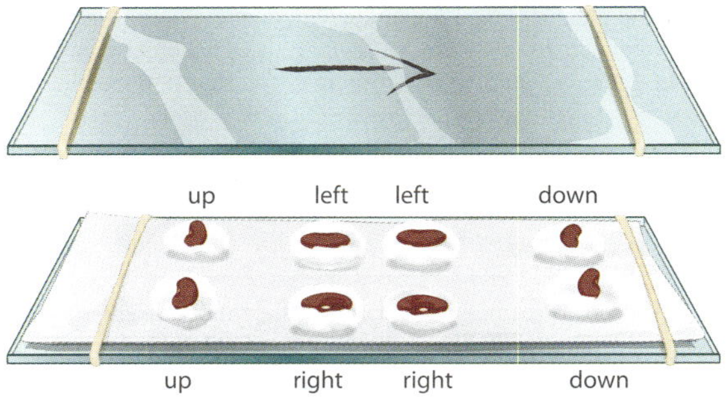

④ Place another glass plate over the top of the bean seeds and secure it with more rubber bands.

⑤ Using a marking pencil, draw an arrow on the top glass plate.

⑥ Set this assembly, with the arrow pointing up, in a dark, warm area.

⑦ Using an eyedropper, add water to the assembly daily and observe the bean seeds daily for germination.

⑧ Once the seeds begin to grow, record their growth patterns over the course of four days in Table 1. For each pair of bean seeds (those facing up, facing right, facing left, and facing down), record what directions the roots grow and what directions the stems grow by marking an X in the appropriate column of the chart.

5. Did seed orientation appear to have any effect on root and stem growth direction? Explain.

No. Roots grow downward and stems grow upward.

6. Do these results support your hypothesis?

Answers will vary.

7. Explain your results in terms of tropism.

The stems exhibited negative gravitropism, and the roots

exhibited positive gravitropism.

ASSEMBLY ORIENTATION ✓

The plate assembly should *not* be flat. It should rather be on its side with the arrow pointing up.

8. Some plants, like crocuses, are planted as underground stems called *bulbs* or *corms*. When you plant a bulb or corm, you are supposed to plant it with one end up and one end down. On the basis of what you have learned in this activity, do you think that the orientation of the bulb matters?

See TE margin for answer.

____

____

____

____

9. Would it be possible to design an experiment that tests for phototropism in the absence of gravitropism?

Answers will vary. Some students may think it impossible, while

others may mention that it *is* possible in space and that it has

been done on the International Space Station.

Table 1

Direction of Growth

Date		Upward-facing beans	Right-facing beans	Left-facing beans	Downward-facing beans
	Roots	☐ up ☐ left ☐ down ☐ right	☐ up ☐ left ☐ down ☐ right	☐ up ☐ left ☐ down ☐ right	☐ up ☐ left ☐ down ☐ right
	Stems	☐ up ☐ left ☐ down ☐ right	☐ up ☐ left ☐ down ☐ right	☐ up ☐ left ☐ down ☐ right	☐ up ☐ left ☐ down ☐ right
	Roots	☐ up ☐ left ☐ down ☐ right	☐ up ☐ left ☐ down ☐ right	☐ up ☐ left ☐ down ☐ right	☐ up ☐ left ☐ down ☐ right
	Stems	☐ up ☐ left ☐ down ☐ right	☐ up ☐ left ☐ down ☐ right	☐ up ☐ left ☐ down ☐ right	☐ up ☐ left ☐ down ☐ right
	Roots	☐ up ☐ left ☐ down ☐ right	☐ up ☐ left ☐ down ☐ right	☐ up ☐ left ☐ down ☐ right	☐ up ☐ left ☐ down ☐ right
	Stems	☐ up ☐ left ☐ down ☐ right	☐ up ☐ left ☐ down ☐ right	☐ up ☐ left ☐ down ☐ right	☐ up ☐ left ☐ down ☐ right
	Roots	☐ up ☐ left ☐ down ☐ right	☐ up ☐ left ☐ down ☐ right	☐ up ☐ left ☐ down ☐ right	☐ up ☐ left ☐ down ☐ right
	Stems	☐ up ☐ left ☐ down ☐ right	☐ up ☐ left ☐ down ☐ right	☐ up ☐ left ☐ down ☐ right	☐ up ☐ left ☐ down ☐ right

---

TABLE 1 

Students may find this table a little challenging to navigate. You may want to help them get started. For each group of seeds, they should indicate in which direction both the roots and stems are growing. In each box, they should check only one of the four choices.

Recording data should not begin until seeds have sprouted and roots and stems are visible.

GERMINATING SEEDS 

Some seeds may not germinate. If only one or two in a single assembly do not germinate, the student should ignore that seed. If three or more seeds do not germinate, the student can make his observations from another student's assembly.

LAB 9D

OLD FARMERS' TALE?

Factors That Affect Germination

IS THERE A WAY TO INCREASE THE GERMINATION RATE OF SEEDS?

Have you ever planted a garden? If so, you know how exciting it is when the seeds you have planted start to germinate and push their way up through the soil. If you search the Internet, you will find lots of different ways that are supposed to help seeds germinate. Some of them may really work while others may not. In this exercise, you and your classmates will test some of these methods.

Procedure

1. Which method for increasing germination listed in the box below do you think will result in the greatest amount of seed germination? Write your answer as a hypothesis.

   Answers will vary. Example: Soaking seeds in fertilizer will result in more of the seeds germinating.

① Obtain 12 seeds from your teacher.

② Choose one of the methods to test.

NAME:

DATE:

Key Question

- How will different factors affect the ability of seeds to sprout and grow?

Equipment

watermelon seeds (12)

20% w/v polyethylene glycol solution, 500 mL

fertilizer solution, 500 mL

sandpaper

refrigerator

water

72-cell seed starting tray

potting soil

labels (6)

bowls (3)

Methods of Increasing Seed Germination

Fertilizer. Soak the seeds in a fertilizer solution for 24 hours.

Polyethylene glycol (PEG). Soak the seeds in a 20% w/v PEG solution for 24 hours.

Refrigeration. Place the seeds in a refrigerator for 24 hours.

Sandpaper. Briefly rub each seed with sandpaper.

Water. Soak the seeds in water for 24 hours.

OLD FARMERS' TALE? 107

LAB 9D OBJECTIVE

- Identify the effects that various factors have on the ability of seeds to sprout and grow.

CONCURRENT LABS 

See the Concurrent Labs teacher note on page 103.

LAB PREP 

Any type of seed will work for this activity. Watermelon seeds are listed here because they are relatively large and germinate quickly. It's best to avoid very small seeds like radish seeds and seeds such as bean seeds that take two or more weeks to germinate.

Divide your class into five groups. Each group should test a different method. If you have a small class, each person can test one of the factors listed. You will need to plant twelve untreated seeds to form a control. If you are using a standard seventy-two-seed starting tray, you will have room to plant all the seeds in one tray (i.e., one tray per class).

Polyethylene glycol (PEG) is a common laxative available at most pharmacies. To make a 20% w/v PEG solution, dissolve 100 g of PEG in 250 mL of water. Then dilute the solution with enough water to make 500 mL of solution.

Any type of commonly available fertilizer will work. Make 500 mL of solution according to the instructions on the package.

NOTES

All the groups will plant their seeds at the same time. Your teacher will plant 12 seeds that were not treated with any method. After all the seeds have been planted, place the plant tray in a warm, sunny spot.

2. In this experiment, what is the group of untreated seeds that your teacher is planting?

the control group

3. After treating the seeds with the method you chose, plant the seeds in a large seed-starting tray about 1.3 cm deep in the soil. Label the rows of seeds with the method used.

4. Throughout the experiment, water the seeds enough to keep the soil moist but not too wet.

5. Each day for 14 days, inspect the tray for germinating seeds. Record in Table 1 the number of sprouts visible in each group for that day.

SAMPLE DATA

The data in Tables 1 and 2 are sample data. Your students' data may be different.

Table 1

Methods of Inducing Germination

Days after planting	Control	Fertilizer	PEG	Refrigerator	Sandpaper	Water
1						
2						
3	0	0	0	0	0	0
4	0	0	0	0	0	0
5	0	0	0	0	0	0
6	0	0	0	0	0	0
7	0	2	0	0	0	0
8						
9						
10	0	3	0	0	2	0
11	2	4	3	0	5	0
12	4	6	7	2	7	1
13	7	9	8	3	10	4
14	8	10	9	6	12	6

3. Why is it necessary to have a control group?

Some or all of the methods used might have actually decreased the seeds' ability to germinate. A control group gives a reference point for comparison with the other groups.

4. Is it better to plant all the seeds in a single tray or in separate trays? Explain.

It is better to plant them in a single tray. Planting them in separate trays would introduce an additional variable. This would be likely to occur when watering the seeds since the amounts of water applied to the different trays would likely vary unless the amounts of water were measured.

6. Rank the groups according to the time it took for the first sprout of each group to appear, with the group that sprouted most quickly listed first. Record your rankings in the *By Sprout Order* column of Table 2.

7. Rank the groups according to the number of seeds that germinate by the end of the experiment. Record your rankings in the *By Count* column of Table 2.

5. Which group(s), if any, sprouted more quickly than the control group?

Answers will vary.

6. Which group(s), if any, had two or more seeds more than the control group sprout?

Answers will vary.

7. Were there any groups with results that were similar to the control group? If so, which one(s)?

Answers will vary.

8. Do the results of this experiment support your hypothesis? Explain.

Answers will vary.

Going Further

9. Some of the methods of increasing germination may work with some types of seeds but not with others. Describe an experiment to test this possibility.

Answers will vary. Students should describe an experiment in which some or all of the methods from this experiment are applied to several different types of seeds. These should then be planted with control groups.

Table 2
Rankings

	By Sprout Order	By Count
1st	fertilizer	sandpaper
2nd	sandpaper	fertilizer
3rd	PEG	PEG
4th	control	control
5th	fridge	fridge
6th	water	water

ANIMAL CHARACTERISTICS

NAME:

DATE:

Are the organisms on the right animals? The ocelot is obviously an animal. The other two organisms look similar to each other, but neither looks like an animal. In spite of that, the purple organism is indeed an animal. The other is a fungus.

On page 193 of your textbook you learned the characteristics of animals. These attributes are what place the sea anemone—the purple organism—and the ocelot in the same kingdom but classify the stinkhorn fungus in a different one.

Both the ocelot and the sea anemone are made of eukaryotic cells, consume their food, have tissues, and can move, though only occasionally and slowly in the case of the sea anemone.

In this activity, you will use the characteristics of animals to distinguish between animals and nonanimals and between different types of animals.

Why **Not** *an Animal?*

In Questions 1–7, indicate which of the animal characteristics from the list on the right each nonanimal lacks.

1. oak tree

   b, d, e

2. *Lactobacillus acidophilus* bacterium

   a, c, d, e, f

3. jack-o'-lantern mushroom

   d, e

4. amoeba

   c, f

5. algal pond scum

   b, c, d, e

6. imperfect fungus

   d, e, f

7. photosynthetic cyanobacterium

   a, b, c, d, e, f

a. is made of eukaryotic cells

b. is a consumer

c. has specialized cells

d. is capable of movement

e. eats food (doesn't absorb food)

f. reproduces sexually

- List the major characteristics common to all animals.

- Explain how the characteristics of animals can be used to classify them.

NOTES

Classification Criteria

8. Does an African lion display bilateral symmetry, radial symmetry, or asymmetry?

   bilateral symmetry

9. Does a planarian have a complete or incomplete digestive tract?

   incomplete digestive tract

10. Does a honeybee have cephalization?

    yes

11. Is a stag beetle (at left) a vertebrate or an invertebrate?

    invertebrate

12. Is a wolf an endotherm or an ectotherm?

    endotherm

REVIEW 10B

IDENTIFYING INVERTEBRATES

NAME:

DATE:

You can probably tell that the animal below is a spider. Your recognition of that fact was an act of classification. You used characteristics like the animal's size, shape, and jointed legs to place it into a group of organisms called *spiders*.

This daring jumping spider captures its prey by hunting rather than by using a web like most spiders do. It poses no threat to humans; in fact, it is less than an inch long.

REVIEW 10B OBJECTIVE

- Compare the major invertebrate phyla.

In the activity below, use the same process to identify the group being described in each question. Use the information on pages 200–204 of your textbook.

Write the common name for the group of animals described.

1. marine animals that have stinging cells to paralyze and capture prey

   cnidarians

2. animals that have soft bodies and are divided into repeating body segments

   segmented worms

3. animals that have spiny skins and live in the ocean

   echinoderms

4. animals that have an exoskeleton and jointed appendages

   arthropods

5. animals that have soft and flattened bodies

   flatworms

6. animals that have soft bodies and, frequently, shells

   mollusks

7. animals that lack symmetry and cephalization

   sponges

8. animals that have complete guts but are otherwise very simple

   roundworms

9. arthropods that have six legs and three body segments

   insects

10. mollusks that have multiple tentacles

    cephalopods

11. arthropods that have eight or ten legs

    crustaceans

12. mollusks that have two-piece, hinged shells

    bivalves

13. arthropods that have eight legs and two body segments

    chelicerates

NOTES

REVIEW 10C

IDENTIFYING VERTEBRATES

NAME:

DATE:

A house cat is obviously a cat. But how would you classify the animal below? It's a South American mammal called a *coatimundi*. It is classified in the same family as raccoons.

Use the information on pages 205–12 of your textbook to identify the groups of vertebrates described below.

1. vertebrates that have feathers and lack teeth

   birds

2. vertebrates that have fur or hair

   mammals

3. smooth-skinned vertebrates that typically live in water when young but can live on land as adults

   amphibians

4. vertebrates that have scales and lungs

   reptiles

5. vertebrates that have gills and bony skeletons and are usually covered by scales

   bony fish

6. ectothermic aquatic vertebrates that have gills and fins

   fish

7. aquatic vertebrates that have gills but lack scales and paired fins

   jawless fish

8. aquatic vertebrates that have jaws, scales, and endoskeletons made of cartilage

   cartilaginous fish

9. animals that live in water as young and on land as tailless adults

   frogs and toads

10. tailed vertebrates that have moist, scaleless skin

    salamanders

11. scaly ectothermic vertebrates that molt their skin

    squamates

12. vertebrates that have scutes and teeth

    crocodilians

13. vertebrates that have two hard, bony shells

    turtles

14. endothermic vertebrates that are connected to their developing young by umbilical cords

    placental mammals

15. furred vertebrates that lay eggs

    monotremes or egg-laying mammals

16. endothermic vertebrates that have pouches

    marsupials

REVIEW 10C OBJECTIVE

- Compare the major vertebrate classes.

NOTES

LAB 10D

A PLACE FOR EVERYTHING

Inquiring into Classification

NAME:

DATE:

Key Questions 

- Why do people classify?
- How does worldview affect classification?

Equipment
none

HOW DOES SOMEONE CHOOSE CRITERIA FOR CLASSIFICATION?

People classify things in order to make general statements about them. We *could* say that it's important to eat iceberg lettuce, spinach, kale, cabbage, collards, romain lettuce, mustard greens, and Swiss chard. But it would be much easier to simply say that it's important to eat green leafy vegetables.

You can often classify a group of objects in multiple ways. For instance, you could classify vegetables by their edible plant part, such as their pods (peas and beans), their leaves (lettuce and cabbage), their stems (celery), their roots (onions and potatoes), and their flowers (broccoli and cauliflower). Or you could classify them by color or by the time of year when they grow best. None of these methods of classification is wrong, but some are probably more workable than others.

In this activity, you will choose a group of objects (anything but vegetables!) and classify them in a way that is workable and allows you to make generalizations.

Procedure

1. Decide with your lab team on a group of objects to classify.

2. Submit your group of objects to your teacher for approval.

3. List the individual members of the group that you plan to classify on a separate sheet of paper.

4. Brainstorm with your lab team to identify characteristics that might be used to classify the objects in your chosen group.

   1. What characteristics will your lab team use to classify objects in your group?

      Answers will vary. Example: sports—

      location, objects and gear used

How would you classify these vegetables?

LAB 10D OBJECTIVES 

- Explain why people classify.

- Indicate how worldview affects classification. **BWS**

MANAGING INQUIRY LABS 

An inquiry lab allows students a degree of freedom, but the teacher should guide the process to achieve the educational goals. This process of guiding students is one of the challenges of inquiry labs. The best method is to think through the goals that you want students to achieve and then use questions to guide them to reach those goals. You want to balance their freedom to investigate the question in the manner they choose with the need to meet educational goals.

PLANNING

Each lab team starts by choosing a group of objects to classify. Teams have been given great freedom to choose their objects. Some possibilities are vehicles, sports, pasta, nutrients, shoes, and jewelry.

You probably want to avoid living organisms. Most attempts at classifying living organisms will probably end up with the students mimicking the Linnaean system rather than creating their own. Attempts to classify people may easily lead into inappropriate classifications.

Make sure that students choose a group of items to classify that contains at least ten items. Groups with fifteen to twenty items are best.

NOTES

 In Area 1, use your chosen characteristics to create a hierarchy chart to classify your objects. If you need help creating a hierarchy chart, see Appendix B of your textbook.

ANSWERS 

The suggested answers in Area 1 use summer Olympic sports (as of 2018) as the group that is being classified. Most of your students will probably choose a different group.

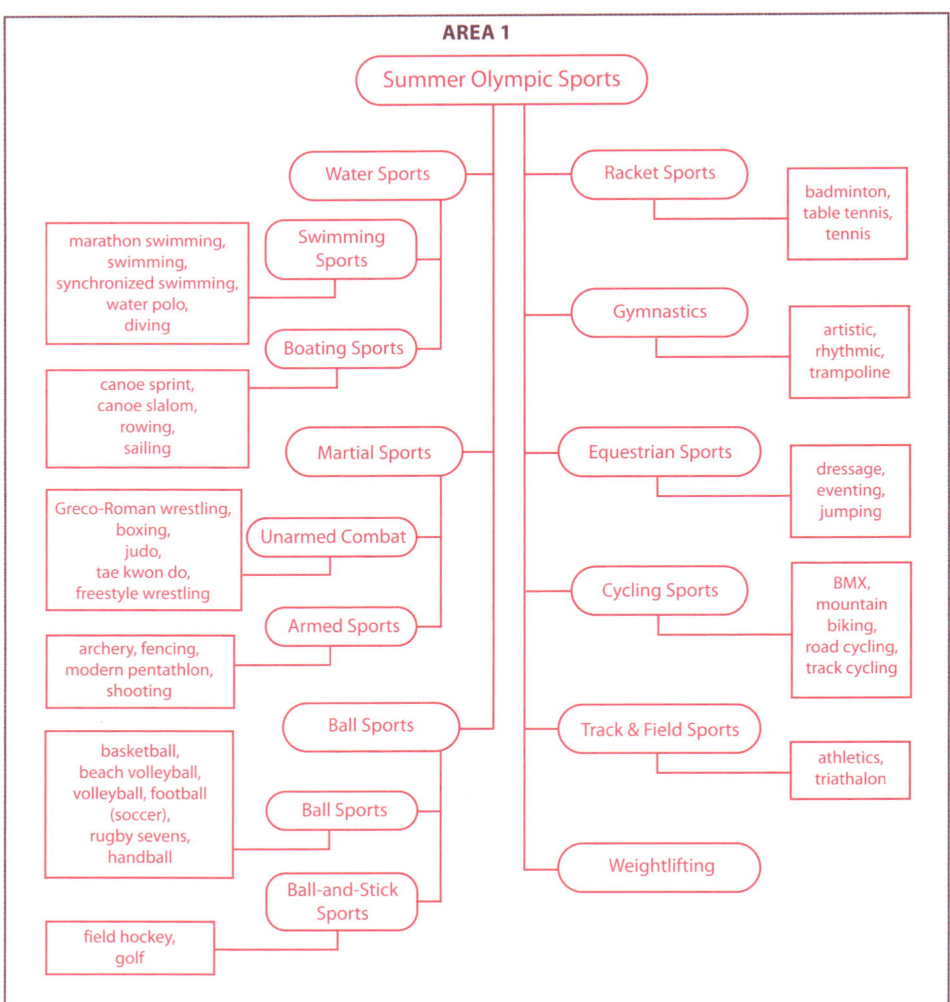

Question 2 Answer

Answers will vary. Examples: Modern pentathlon could go in Swimming Sports, Track and Field Sports, or Equestrian Sports. Table tennis and tennis could go in Racket Sports or Ball-and-Stick Sports.

2. Are there any individual items in your group that could fit into more than one of your categories? Explain.

See TE margin for answer.

3. Are there any categories that you chose not to use in your classification system? If so, what are they?

See TE margin for answer.

4. If you listed some categories in Question 3, why did you choose not to include them in your classification system?

Answers will vary. Example: It seemed more natural not to include them.

Going Further

Scientists who work to classify living organisms sometimes struggle to group species together. Part of the reason for this is that scientists debate what criteria to use in classification. Some evolutionary scientists have suggested classifying marine mammals in the order Cetartiodactyla, a combination of most hoofed mammals and marine mammals, such as whales.

5. Would classifying hoofed mammals and marine mammals as a group separate from other mammals be useful in making general statements about these animals? Explain.

No. The mammals in the two orders are so different structurally and have such different lifestyles that classifying them together would not be useful in generalizing about them.

6. How would a person's worldview affect whether he groups hoofed mammals and marine mammals together?

A creationist will see them as separate groups of animals created by God, so he would see no need to group them together unless doing so helped in generalizing about them. To use biblical terminology, they are two different kinds (Gen. 1:25). An evolutionist believes that all organisms are descended from a common ancestor, so he will place these two groups together if he thinks they are closely related.

Should these animals be classified together in one order?

Question 3 Answer

Answers will vary. Examples: whether the game involves a net, team versus individual sports, contact versus limited contact versus noncontact sports

A DIFFERENT PERSPECTIVE

Discuss with students that people with different worldviews are not being disingenuous. They are sincere. If we want to convince someone of our views, we need to address his presupposed meta-narrative.

LAB 10E

RECORDS

Scientific Journaling

NAME:

DATE:

HOW CAN I RECORD SCIENTIFIC OBSERVATIONS?

Most people have family pictures in their homes. They show families at different stages of their lives. They may help the family remember relatives who live in different places or have passed away.

Scientists who explore new areas often keep journals for the same reason. They need to remember all the details about the organisms they discover, especially since some of them may be new to science. They will try to collect samples of the organisms that they discover, but there are many details that a sample can't reveal. Scientists can't rely on their memories to remember these details, so they often record them in journals. A scientist's journal will record his observations, such as the kind of habitat an organism lives in and where specimens were seen or collected.

In this activity, you will create a journal much as a scientist would. But instead of collecting samples, you will take pictures of the organisms that you journal.

Key Questions

- How do I identify organisms?
- How do I journal about organisms?

Equipment

camera

small notebook

field guides

LAB 10E OBJECTIVES

- Identify organisms using field guides and dichotomous keys.

- Gain experience in journaling about organisms.

LAB SCHEDULING

If you use this laboratory activity, you may want to do so before Chapter 10 is studied since your class will likely be completing that chapter during the winter months.

You may direct students to focus their journaling on a specific phylum or class of animals, such as birds or insects. Or you may have them focus on any animals.

This lab activity could be assigned as homework or could be part of a field trip to a park, field, or forest area.

Ideas for Locating Invertebrates

1. Look under stones, boards, and logs.
2. Dig up and turn over a shovelful of earth. Watch closely and capture any invertebrates that scurry away.
3. Check around outdoor lights at night.
4. At night, put a light over a tub of water that contains a spoonful of kerosene. In the morning, collect the invertebrates that came overnight.
5. Leave an open sandwich outside for an hour or two. Invertebrates will soon be attracted to the food.
6. Use an insect net to capture flying insects. Disturbing bushes and tall grass will often arouse many flying insects.
7. Check pool filters for beetles and other large invertebrates that have fallen into the water.

RECORDS **121**

NOTES

Procedure

1. In your notebook, write down where you are observing animals. Write down the date, the time of day you are making observations, and the weather conditions.

2. Find animals by either sitting quietly or by actively searching for them. Vertebrates are often large enough to be easily seen. You may have to hunt to find invertebrates. See the Ideas box on page 121 for ideas.

3. Take pictures of the animals you find. You should try to find at least twelve different animals, at least half of them invertebrates.

4. Number each picture file and write down each specimen's number in your notebook.

5. Use dichotomous keys, field guides, or apps to identify each animal in your notebook as precisely as possible. For many animals, you may not be able to identify the species, but you should at least identify the order each one belongs in.

6. Use the information from your notebook to complete Table 1.

7. Submit your notebook and pictures to your teacher.

1. How many of the animals you identified are invertebrates?

   Answers will vary.

2. What percentage of animals you identified are invertebrates? Divide the number you wrote in Question 1 by the total number of animals you identified and multiply by 100%.

   Answers will vary.

3. Invertebrates make up 97% of the world's animal species. Was your answer in Question 2 close to this number?

   Answers will vary but will most likely be "No."

4. If you answered "No" to Question 3, why do you think the percentage of invertebrates in your collection is different from the percentage of invertebrates worldwide?

   Answers will vary. Students should point out that vertebrates are larger than invertebrates and so are easier to see. They may also mention the time of year that it is—if this activity is done in the late fall or winter, endothermic vertebrates are more likely to be active than invertebrates or ectothermic vertebrates.

Grouping Animals

1 Using the information from Table 1, complete Column 1 of Table 2 by identifying the classes of animals you collected.

2 Complete Column 2 by writing the number of specimens for each class in Column 1.

3 For each row of Table 2, complete Column 3 by dividing the number of specimens in Column 2 by the total number of specimens in your collection.

4 For each row of Table 2, do some research and complete Column 4 by finding the percentage of species of animals worldwide in each class you identified.

5. Were any classes overrepresented in your collection? If so, which ones?

   Answers will vary.
   ____

6. Were any classes underrepresented in your collection? If so, which ones?

   Answers will vary.
   ____

7. Suggest reasons why classes were overrepresented or underrepresented in your collection, if any.

   Answers will vary. Possible reasons may include size, coloration, and activity during the time of day or year.
   ____
   ____
   ____

124 LAB 10E

Table 1

Animals Identified

Specimen number	Order	Family (if known)	Genus (if known)	Species (if known)
1				
2				
3				
4				
5				
6				
7				
8				
9				
10				
11				
12				

Table 2

Percentages of Classes Identified

Column 1 Class identification	Column 2 Number of specimens	Column 3 Percent of collection	Column 4 Percent of animal kingdom

DIGESTION CONCEPT MAP

NAME:

DATE:

Create a concept map (see Appendix B in your textbook) that includes all the terms listed below. You may use additional terms, examples, or labels if necessary. As you think about how to build your concept map, think about the following questions: What do animals need energy for? What do they obtain it from? How do they extract it from food?

carnivore	gizzard	omnivore
chemical digestion	herbivore	prey
digestion	homeostasis	stomach
filter feeder	intestines	teeth
food	mechanical digestion	

The example shown is one solution. Other solutions are possible.

REVIEW 11A OBJECTIVE

- Compare various ways that animals obtain energy.

CONCEPT MAPS 

Students who need assistance may be referred to Appendix B in *Life Science* Student Edition 5th Edition.

HOW DO I GRADE THIS?

There is not just one way for students to achieve the desired outcome of this activity. When grading, check to see that the student has included the required terms and shown the correct relationships between them. You may assign partial credit according to the degree to which they have accomplished those two tasks.

NOTES

ANIMAL CIRCULATION

NAME:

DATE:

Label the atria and ventricles in the diagrams below. Use colored pencils to show the pathway for oxygenated blood (red) and deoxygenated blood (blue). Draw arrows to show the direction of blood flow.

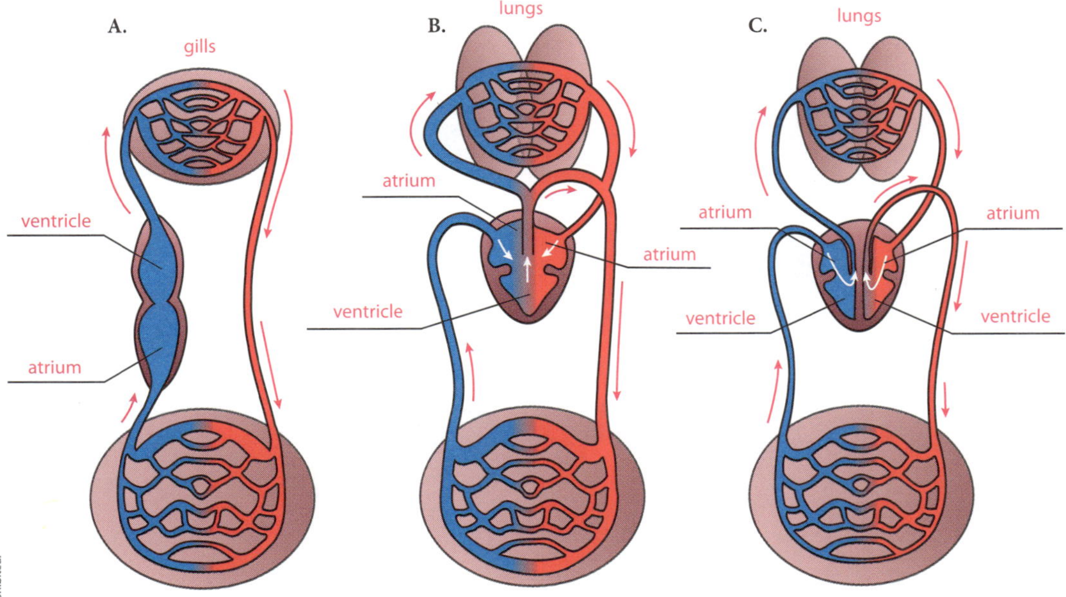

1. Are the circulatory systems shown above open or closed? Explain.

   closed (The blood always remains inside vessels.)

In Questions 2–8, tell which type of circulatory system above (A, B, or C) you would find in each animal.

B **2.** iguana

C **3.** alligator

A **4.** mosquitofish

C **5.** opossum

B **6.** corn snake

A **7.** mako shark

C **8.** house sparrow

REVIEW 11B OBJECTIVES

- Distinguish between open and closed circulatory systems.

- Compare the circulatory systems of different vertebrate groups.

NOTES

REVIEW 11C

SUPPORT, MOVEMENT, AND CONTROL

NAME:

DATE:

Read the phrases below and then decide which animal or animals of the ones shown below is (are) described by each phrase. The phrase may describe none of the animals.

REVIEW 11C OBJECTIVES

- Compare types of animal support.

- Compare types of animal movement.

a. beetle

b. crab

c. jellyfish

d. pigeon

e. squid

f. all of these

g. none of these

_a, b_   **1.** has an exoskeleton

_c_   **2.** has a nerve net with no brain

_c, e_   **3.** moves using jet propulsion

_d_   **4.** has bones joined together by ligaments

_g_   **5.** has no nervous system

_g_   **6.** is sessile

_a, b, d, e_   **7.** has a brain to process nerve impulses

_c, e_   **8.** has a hydrostatic skeleton

_d_   **9.** has an endoskeleton made of bone

_f_   **10.** is capable of locomotion

_c, e_   **11.** relies on water pressure for support

_g_   **12.** has an endoskeleton made of cartilage

_f_   **13.** has a muscular system

_d_   **14.** has a brain and spinal cord

NOTES

STAYING COZY

Conserving Heat: Wool Versus Down

WHICH IS A BETTER INSULATOR, WOOL OR DOWN?

On a cold winter day, you pick your coziest clothes to stay warm. Similarly, endothermic animals like mammals and birds need to conserve the heat energy that their bodies produce to maintain homeostasis. We call materials that perform this function *insulators*. Mammals have fur to insulate their bodies. Bird bodies are insulated by special feathers called *down*. In this activity, you'll investigate whether one of these is a better insulator than the other.

NAME:

DATE:

Key Questions

- How can the abilities of different materials to conserve heat energy be compared?
- Is one kind of body covering better than another at conserving heat?

Equipment

containers (cup or glass), each with a lid that has a hole (3)

unspun wool

down

thermometers (3)

freezer

LAB 11D OBJECTIVE

- Compare the effectiveness of wool (fur) and down (feathers) as insulators.

OOPS, WRONG CUP! 

Double-walled plastic tumblers with lids and straws are commonplace today. These are *not* suitable containers for this lab activity since they are designed to *prevent* heat loss. Glass containers are best for this activity.

WHICH WOOL? 

Unspun wool, such as batting or roving, is best for this activity.

BLUBBER 

As students think about materials to keep animals warm, they may ask about the effectiveness of blubber. Have students test this by using shortening to simulate blubber.

CONTROLLED EXPERIMENT 

Help students think about experimental design. As they "fill" the containers, do they want to use equal volumes of material or equal masses?

1. Which do you think will do a better job of conserving heat energy, wool or down? State your answer in the form of a hypothesis.

   Answers will vary.

Procedure

1. Fill one container with wool and another with down. The containers should have equal amounts of the materials to be tested. Label the containers *Wool* and *Down*. Label a third empty container *Control*.

2. What is the purpose of the control container?

   The purpose of the control container is to enable the student to more accurately compare the effect of the wool and down on temperature change.

NOTES

2 Put a lid on each container and insert a thermometer through the hole in each lid. The thermometer should be positioned so that the thermometer bulb is about midway in the container. Do not allow the thermometer bulb to touch the bottom or side of the container.

3 Record the starting temperatures of all three containers in Table 1 to the nearest tenth of 1 °C. Then put all three in the freezer. (*Note*: All three should be at approximately the same starting temperature.)

4 At five-minute intervals, record the temperatures of all three containers in Table 1 to the nearest tenth of 1 °C.

3. What was the experimental variable in this experiment?

the type of material (wool or down)

4. Which container had the greatest temperature change during a five-minute interval?

the control container

5. Which container had the greatest temperature change from start to finish?

the control container

6. Which container had the smallest temperature change from start to finish?

Answers will vary. Either the wool or the down container will

have the smallest change.

Summing Up

7. Was your hypothesis correct?

Answers will vary.

8. Are wool and down insulators? Explain.

Yes. The temperatures in the filled containers did not drop as

much as that of the control container.

Going Further

9. What do you think would have happened if you had left the lid off the control container?

The container would have cooled more quickly.

10. If you had left the lid off the control container and put the lids on the two experimental containers, could you have reached a reliable conclusion? Explain.

No. In a controlled experiment, there must be only one difference among the groups. In this experiment, that difference was the type of material. Leaving the lid off would introduce a second variable into the experiment.

11. Explain how wool and down could be a disadvantage for endothermic animals.

In hot climates, endothermic animals may need to lose body heat rather than conserve it.

Table 1

Temperature Data

Time	Control (°C)	Wool (°C)	Down (°C)
At start	20	20	20
At 5 minutes	12	17	16
At 10 minutes	6	10	11
At 15 minutes	4	6	7
At 20 minutes	3	4	5
At 25 minutes	2	3	3

COOL IT!

Of course, mammals and birds *do* live in deserts and must often rid themselves of excess heat energy. They can do this through a variety of means, including sweating, panting, or taking a dip in a local water source.

GOING FURTHER 

You can expand this activity by testing different kinds of feathers (e.g., flight feathers versus down) or different kinds of wool (spun versus unspun, sheep versus alpaca).

LAB 11E

CRANK UP THE HEAT

The Effect of Temperature on Fish Respiration

HOW DOES CHANGE IN TEMPERATURE AFFECT A FISH'S RATE OF RESPIRATION?

A run up a flight of steps or a good workout can leave you huffing and puffing for air. You need more oxygen when you are physically active. Believe it or not, fish are the same way! They breathe, or respire, faster when they are more active. Other factors can affect a fish's respiration rate too. For instance, fish respire faster when they experience stress. In this lab activity, you'll conduct an experiment to determine whether temperature affects a fish's respiration rate.

1. What do you think—will fish respire faster in warmer water or cooler water? State your answer in the form of a hypothesis.

   Answers will vary.

Procedure

❶ Place a live fish in a small aquarium. The water in the aquarium should be the same as the water to which the fish is accustomed. Do not use tap water unless it has been dechlorinated. Place a thermometer in the aquarium. Allow the fish several minutes to recover from the stress of being moved.

NAME:

DATE:

Key Questions 

- How can the rate of a fish's respiration be measured?
- Do fish respire faster in warmer water or cooler water?

Equipment E

live fish

small aquarium

thermometer or temperature probe

plastic bag

hot water

ice cubes

LAB 11E OBJECTIVES

- Observe the respiration of a fish.
- Determine the effects of temperature on the respiration rate of a fish.

DEMONSTRATION 

You may want to do this lab activity as a demonstration. If so, have several students at a time observing the operculum movements.

You should probably try this experiment a few days in advance on the type of fish you are going to use. Some fish may show very little change within a 10° span. You may need to extend the temperature changes a few degrees.

Be prepared to do other things while you are waiting for the water to change temperatures. By having a student near the aquarium to monitor the temperature, you can continue to lecture or do other activities while waiting.

DISSOLVED O$_2$ AND WATER TEMPERATURE 

Some students might be aware that the dissolved oxygen content of water is greater at lower temperatures. They may be inclined to hypothesize that the fish will respire less at lower temperatures because more oxygen is available. While it is true that more oxygen is available at lower temperatures, a fish's metabolism is also slower at lower temperatures and thus requires less oxygen.

You may choose to have your students do their own research into the link between temperature and the amount of dissolved oxygen in the water. You can then discuss with them the increased potential danger for fish when high temperatures are coupled with increased activity or stress.

CRANK UP THE HEAT **135**

NOTES

A goldfish would be a good (and inexpensive) fish to observe. It should be at least 5–8 cm long. If it is much smaller, students will not be able to see the movements of the operculum easily. If the fish is strong and healthy, it will recover from these temperature changes; if not, it may suffer stress and become weakened.

Use a relatively small aquarium for this investigation. The larger your aquarium is, the longer it will take for the water to heat and cool. A 3.8 L glass or plastic tank about three-fourths full works well.

The best water source for this experiment is water from the fish's usual aquarium. Well water, if used, should be allowed to stand for 24 hours to allow dissolved gases to gas off.

VISUALIZING DATA

Some students will benefit from a visual representation of their data. Consider having students, especially visual learners, create a bar or line graph of their data.

2. Why do you need to wait for the fish to recover before measuring its respiration rate?

Stresses like moving a fish can affect its respiration rate, and this could affect the results of the experiment if the fish is not allowed to recover before collecting respiration data.

❷ Locate and observe the operculum, the flap that covers the fish's gills. The fish moves this flap back and forth to cause water to circulate over its gills. In the gills, gases are exchanged between the water and the blood.

❸ Carefully count the number of times the operculum beats in 15 seconds. Multiply this number by four. Repeat the observation three more times. Record your findings in the *Room Temperature* row of Table 1.

3. Why do you need to multiply the number of times the operculum beats in 15 seconds by 4?

This number represents how many times the operculum beats in one minute.

4. Why do you need to measure the fish's respiration rate multiple times?

This reduces the effect of error on the observations.

❹ Place several cups of hot water in the plastic bag and close it tightly. Slowly and without disturbing the fish, lower the bag into the aquarium. Do not open the bag or stir the water in the aquarium. Wait for the aquarium water to increase 5° above the original temperature. Then slowly remove the bag of hot water. Repeat your observations of the operculum's movements and record your findings in the *Warm* row of Table 1.

Table 1
Fish Respiration Data

Temperature of water	Number of beats in one minute				
	First count	Second count	Third count	Fourth count	Average beats per minute
Room temperature	72	80	64	56	68
Warm	80	68	88	84	80
Cold	56	72	68	60	64

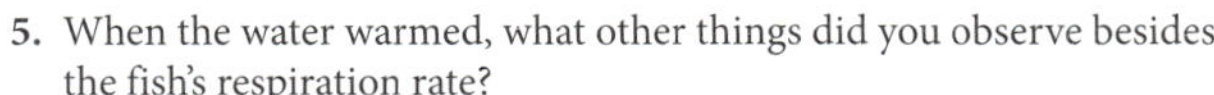

5. When the water warmed, what other things did you observe besides the fish's respiration rate?

Answers will vary. Fish may move to a cooler area of the tank to relieve the stress of changing temperature conditions or they may increase their rate of activity.

5 Place several ice cubes in the plastic bag and close it tightly. Slowly and without disturbing the fish, lower the bag into the aquarium. Do not dump the ice cubes into the water or stir the water in the aquarium. Wait for the aquarium water to cool 5° below the original temperature. You may need to replace the original ice to obtain the desired temperature change. Afterward, slowly remove the bag of ice. Repeat your observations of the operculum's movements, and record your findings in the *Cold* row of Table 1.

6. What other things, besides the fish's respiration rate, did you observe when the water cooled?

Answers will vary. Fish may rest on the bottom of the tank or reduce their rate of activity.

6 Record the average for each row in the *Average beats per minute* column.

7 Allow the aquarium water to gradually return to its normal temperature.

Summing Up

7. At what temperature (room, warm, or cold) was the fish's operculum most active?

warm

8. At what temperature was the fish's operculum least active?

cold

9. How did the average rate of operculum beats compare with the amount of activity of the fish?

The fish moved its operculum more as it became more active.

CRANK UP THE HEAT **137**

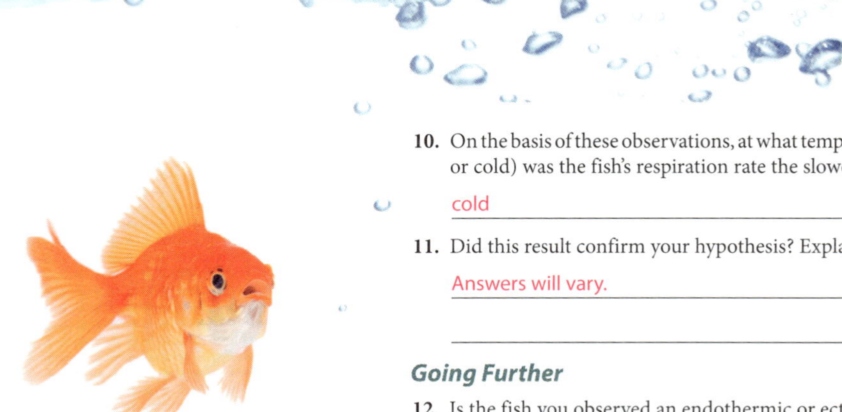

10. On the basis of these observations, at what temperature (room, warm, or cold) was the fish's respiration rate the slowest?

cold

11. Did this result confirm your hypothesis? Explain.

Answers will vary.

Going Further

12. Is the fish you observed an endothermic or ectothermic animal?

ectothermic

13. Explain how the results of this investigation support your answer to Question 12.

The amount of respiration carried on by the fish depended on the temperature of the fish's environment.

14. On the basis of your findings in this activity, how do you think changing temperatures will affect the respiration rate of other ectothermic vertebrates, such as amphibians? Explain.

The respiration rates of other ectothermic vertebrates, such as amphibians, will change as the temperature changes. Their respiration rate will decrease at lower temperatures and increase at higher temperatures.

ANIMAL REPRODUCTION

NAME: _______________

DATE: _______________

Kittens are really cute. They are also key to the survival of the species. When God created land animals on Day 6 of Creation, He gave them the ability to reproduce, resulting in fun creatures like the little fur ball shown at the bottom of the page.

In Questions 1–12, the statements contain information on animal reproduction. In each one, choose the term from the list on the right that correctly completes the statement.

___d___ 1. The process of gametes uniting is called __________.

___k___ 2. Red dace fish build nests called *redds*, where the female lays eggs while the male releases __________.

___e___ 3. Some animals __________ their young while they develop inside an egg.

___g___ 4. Mammals bear live young, and birds lay eggs with hard shells. Both exhibit __________.

___n___ 5. The primary stored food that a developing chick uses to grow is the __________.

___i___ 6. A pig zygote forms when an egg from its mother's __________ is fertilized.

___j___ 7. A baby animal has DNA from both its mother and father. This is because animals reproduce by __________.

___l___ 8. Male gametes are produced in the __________.

___m___ 9. Mammals are different from other animals in that they carry their young in a(n) __________.

___c___ 10. Animals that live on land or lay eggs with hard shells do not exhibit __________.

___f___ 11. Most animals that exhibit __________ hold their eggs inside the female's body until the eggs hatch.

___a___ 12. The gamete produced by a female animal and a zygote surrounded by yolk and a shell are both called __________.

a. eggs

b. external development

c. external fertilization

d. fertilization

e. incubate

f. internal development

g. internal fertilization

h. metamorphosis

i. ovary

j. sexual reproduction

k. sperm

l. testes

m. uterus

n. yolk

13. In the spaces below, write the definitions of the two terms that were not used.

external development—development that takes place outside the mother's body, typically as eggs

metamorphosis—a type of growth that involves a change to the body

REVIEW 12A OBJECTIVES

- Differentiate between external and internal fertilization.

- Contrast internal and external prenatal development.

QUESTION 7: PARTHENOGENESIS

A few students may be aware of parthenogenesis, the process whereby some female animals produce offspring without a male fertilizing an egg. While this does occur, it is fairly rare and is beyond the scope of this course.

ANIMAL REPRODUCTION **139**

NOTES

REVIEW 12B

ANIMAL BEHAVIOR

NAME:

DATE:

You don't usually need to call a pet to dinner. As soon as it hears the food being poured into the bowl or the can opener running, it heads for its food dish. Is this an innate or a learned behavior? According to the information in your textbook, this is a learned behavior.

The examples below are similar to the one above. Read each one and choose the type of behavior being described from the list on the right. One example has more than one answer.

___b___ 1. When given the proper materials, a bird kept isolated for its entire life builds a nest typical of its species.

___c___ 2. During dinner, a dog sits under the chair of the family member who often gives it bits of food.

___b & c___ 3. After watching the older lions in its pride, a lion cub stalks its first prey.

___c___ 4. A monkey uses a long stick to get food that is beyond its reach.

___a___ 5. When a bird becomes cold, its feathers fluff up and it begins to shiver.

___b___ 6. A honeybee collects nectar from flowers to produce honey.

___a___ 7. A honeybee stings your foot when you step on it while walking barefoot across a lawn.

___c___ 8. At a security checkpoint, a trained dog sniffs travelers' bags. When it smells drugs, it barks and claws at the bag.

___c___ 9. After twenty tries, a rat can run a maze in less than one-third of its original time.

___b___ 10. Screeching loudly, a mockingbird swoops down toward you as you walk by the tree where its nest is.

a. innate behavior—reflex

b. innate behavior—instinct

c. learned behavior

REVIEW 12B OBJECTIVES

- Classify animal behaviors as innate or learned.

- Identify the general forms of communication that animals use.

ANIMAL BEHAVIOR **141**

NOTES

LAB 12C

ARE YOU MY MOMMY?

Animal Reproduction Worksheet

ARE AN ANIMAL'S REPRODUCTION STRATEGIES RELATED TO THE NUMBER OF OFFSPRING THAT IT PRODUCES?

Dogs and cats often have six to eight offspring at a time. Elephants and cattle usually have only one. Does size have anything to do with the number of young born at one time?

These kinds of questions are part of determining an animal's *reproductive strategies*. Some animals have few offspring while others have many. Two other aspects of an animal's reproductive strategy include whether it lays eggs or gives birth to live young and whether it cares for its young. In this lab activity, you will see which of these two aspects of reproductive strategies is more closely connected to the number of young that an offspring has.

Procedure

① Using online or printed sources, find the information needed to complete Table 1. Some answers have been provided for you.

NAME:

DATE:

Key Question ?

- How many offspring do animals produce at one time?

Equipment E

reference resources

Excel software

LAB 12C OBJECTIVE

- Compare the number of offspring that various animals reproduce at one time.

GRAPHING 

Students can use an Excel spreadsheet for calculations and the creation of bar graphs. If they don't have access to the software, they can use a calculator for calculating the averages and graphing paper to draw the graphs.

PARENTAL CARE

It's important to remember that reproductive strategies are very complex and that not all parental care is the same. The parental care of a blue crab consists of carrying eggs to the sea and then releasing the larvae into the ocean, after which the mother does nothing for the young. On the other hand, elephants care for their young for several years after birth.

WORLDVIEW AND ANIMAL REPRODUCTION

Evolutionists often discuss reproductive strategies in the context of their evolutionary model. They suggest that an animal that provides a lot of parental care has evolved this behavior to accommodate the fact that it produces few offspring. But the biblical view recognizes that God has created animals with many different reproductive strategies that allow them to survive in their habitats.

ARE YOU MY MOMMY? **143**

NOTES

1. What is the average number of eggs produced by the egg layers in this list?

Answers will vary slightly but should be around 600,000.

2. What is the average number of offspring produced by the live bearers in the list?

Answers will vary slightly but should be around 27.

❷ Use your answers to Questions 1 and 2 to create a bar graph using an Excel spreadsheet.

3. According to your data, do egg layers or live bearers have a higher average number of offspring?

egg layers

4. Is the difference in the lengths of the bars on your graph small or large?

large

5. What is the average number of offspring for the animals that do not provide parental care?

Answers will vary but should be around 650,000.

6. What is the average number of offspring for the animals that provide parental care?

Answers will vary but should be around 230,000.

❸ Use your answers to Questions 5 and 6 to create a bar graph using an Excel spreadsheet.

7. According to your graph, how is parental care related to the number of offspring an animal has?

Animals that do not provide parental care have a greater

average number of offspring.

8. Is the difference in the lengths of the bars on your graph small or large?

large

9. Which graph showed the larger difference between the two bars?

the first graph comparing egg layers to live bearers

10. On the basis of your answer to Question 9, which factor would be a better predictor of the number of offspring that an animal has?

whether an animal is an egg layer or a live bearer

Going Further

11. What are some other differences that might affect the number of off-spring that you could investigate using this data?

Answers will vary. Students may mention the different phyla or classes of animals, vertebrates compared with invertebrates, or endothermic compared with ectothermic.

12. You decide to compare the number of offspring of mammals with that of other animals using the data in Table 1. What would be a problem with this?

Answers will vary. Students should mention that there is only one mammal in Table 1, so there is not enough information to provide any meaningful trends.

Table 1

Animal name	Live bearer or egg layer?	Average number of eggs or young at one time	Parental care provided?
Atlantic surf clam	eggs	6,500,000	no
bald eagle	eggs	2	yes
black widow	eggs	475	yes
blue crab	eggs	2,000,000	yes
brown pelican	eggs	3	yes
bullfrog	eggs	11,500	no
common octopus	eggs	300,000	yes
eastern diamondback rattlesnake	live	12	no
emperor scorpion	live	21	yes
giant katydid	eggs	125	no
guppy	live	55	no
knobbed whelk	eggs	8,000	no
lined seahorse	eggs	825	yes
lobster	eggs	7,500	yes
monarch butterfly	eggs	400	no
prairie dog	live	4	yes
snapping turtle	eggs	30	no
stag beetle	eggs	20	no
Surinam toad	eggs	80	yes
tiger shark	live	45	no

LAB 12D

TRAPPED!

Building a Better Insect Trap

WHAT IS THE BEST WAY TO CATCH A VARIETY OF INSECTS?

Did you know that for every human on the earth, there are 125 million insects? Insects are a huge class; in fact, they're the largest class of animals. It is estimated that there are at least 2 million species of insects out there. *Entomologists*—the scientists who study insects—use different methods to study them. One common way is to trap them.

But some traps work better than others. To design a good insect trap, scientists must understand an insect's behavior. Some insects are attracted to things that other insects are not. Some insects can fly while others cannot. Scientists must research and understand the insects they need to capture in order to design a good trap.

Procedure

Your task for this lab activity is to design an effective insect trap. It must be able to catch insects in such a way that they can still be identified after removal from the trap.

❶ Do some research on insect behavior and things that attract insects.

❷ Your teacher has provided a selection of possible materials to use for building your trap. On the basis of what you have learned about insects and the materials available for you to use, design a trap that can attract and catch multiple types of insects.

❸ Design your trap so that the insects cannot escape once they have entered the trap.

1. How will your insect trap continue to attract insects without releasing those already caught?

   Answers will vary. Some students may use a funnel-shaped

   entrance, a sticky trap, a water trap, or some other method.

NAME:

DATE:

Key Questions 

- What are some good ways of attracting insects?
- How can you trap insects?

Equipment 
teacher-provided materials

LAB 12D OBJECTIVES 

- Investigate ways of attracting insects.
- Demonstrate ways to trap insects.

OBTAINING EQUIPMENT

Most of the equipment listed for this lab activity in the Equipment and Materials list is commonly available at grocery or home improvement stores.

Note that ethylene glycol should not be substituted for the propylene glycol because ethylene glycol is toxic to mammals.

NOTES

4 Now build it! Pull together the materials that you need and construct your insect trap.

5 Take a picture of your insect trap and paste it in Area A.

AREA A

6 Place your insect trap where insects are likely to be and leave it there for 24 hours.

7 Using a field guide or key, identify the orders of insects that you captured.

8 Using your results, fill out the *Trial 1* section of Table 1.

9 Repeat Steps **6** – **8** for Trial 2 and Trial 3 at new locations.

Analysis

2. How effective was your trap design at achieving the goal of catching insects while keeping them in an identifiable state?

Answers will vary.

3. What changes, if any, would you make to your design after evaluating your results?

Answers will vary.

NAME: ______________________________

4. How might designing insect traps be useful?

Answers will vary. Students may mention protecting people from harmful insects that carry disease or collecting insects to study them.

① Research a disease that is caused or spread by an insect.

5. What is the disease and the insect?

Answers will vary. Example: malaria and mosquitos in the genus *Anopheles*

Table 1

Trial	Order of captured insect	Number of each type captured	Total number of captured insects
Trial 1			
Trial 2			

Trial	Order of captured insect	Number of each type captured	Total number of captured insects
Trial 3			

TRAPPED!—BUILDING A BETTER INSECT TRAP

What is the best way to catch a variety of insects?

In this STEM activity, students design and build an insect trap. The activity will guide them through researching insect traps and through the engineering process of design, construction, testing, redesign, and retesting. You will need to provide design parameters and testing procedures.

PREPARATION

Determine the materials you want your students to use for their trap. Then decide on the parameters for the project (see *Plan the Design* below).

1 If the students will work on this project exclusively at school, then gather the materials you want them to use. If they will work on this project away from school, then determine the limitations for types of materials and equipment, such as the following:

a. Allow only the use of household materials not designed for trapping insects.

b. Select a cost limit for any items purchased by the students.

c. Have students get your approval for any item not on your approved list.

2 Divide students into groups of two to four.

3 Select a deployment area for all the groups. Ensure that you have sufficient space to have no traps within 2m of other traps.

PROCEDURE

PLAN THE DESIGN

1 Have students research insect traps and insect lures.

2 Review the instructions for the project with students.

3 Allow students to learn about permissible materials for building their trap. Give them the opportunity to conduct tests with the materials.

4 Each student must draw and label a proposed design.

5 Each group should discuss their findings from materials testing and their individual designs to reach a consensus on the design that they will use.

6 Issue materials and allow groups to construct their traps.

TEST THE DESIGN

1 Have teams install their traps in the location they selected within the deployment area. Testing should span at least an overnight period. After the testing period, the teams will examine their traps, recording how many and what types of insects have been captured.

2 If necessary, the teams should make adjustments in their designs to improve performance.

3 Have teams retest their modified traps. Retesting should be in the same location within the deployment area. The traps should be deployed at the same time of day as the first trial. Leave the traps for the same length of time as before. After the second testing period, teams should examine their traps, recording how many and what types of insects have been captured.

4 The teams should share their results with each other. Each team should break down the results according to lure, kinds of traps, and types and number of insects. They should also compute the average number of insects and types of insects caught in each type of trap.

5 Allow the teams to discuss their results. Each student should then write conclusions describing and justifying their results. Students should also suggest possible improvements to their designs.

REVIEW 13A

HUMANS AND ANIMALS

NAME:

DATE:

Read each Scripture passage, looking for what it teaches us about humans and animals and their relationship to each other. Then fill in Table 1 with what you learn from each passage.

Table 1

Scripture	Humans	Animals
Genesis 1:24–28; 2:19–22	created in the image of God as man and woman; given dominion over animals	created by God "after their kind"; named by man
Genesis 9:2–6	can kill animals for meat; death penalty for killing man	can be killed for meat; death penalty for killing man
Psalm 8:4–9	given glory and honor by God; given dominion over animals	under the dominion of man
Proverbs 12:10	must be kind masters to animals	in the care of man

1. What did you learn about how humans and animals are alike?

   God created them both, and they both deserve death for killing man.

2. What did you learn about how humans and animals are different?

   God created humans in His image and gave them dominion over the animals. God did not create animals in His image, and He placed them under the dominion of man.

REVIEW 13A OBJECTIVE

- Distinguish humans from animals.

HUMANS: AN EVOLUTIONARY VIEW 

Evolutionary biologists view humans as highly evolved primates that evolved from an ancestor common to humans and apes. This view directly contradicts biblical teaching about the creation of man in the image of God. This view also devalues humans by putting them on the same level as animals. Through creation and the Creation Mandate, God clearly placed man in a position superior to all animals.

NOTES

SKIN STRUCTURE

NAME:

DATE:

Oh, dear! You're an aide in Mr. Kvetchmeister's class, and he has asked you to proofread a handout that he is creating for another class. He must not have been paying close attention to his work because there are some mistakes. First, check the labels on the diagram. If a label is correct, leave it as is; if it is incorrect, draw a line through the label and write the correct label beneath it.

Then, for the True/False questions that follow, place a check mark in the answer blank if the question and its suggested answer are correct. If the answer to the question is incorrect, draw a line through the term in the statement that makes it incorrect and write the correct replacement term in the answer blank.

REVIEW 13B OBJECTIVES

- Describe the layers of human skin.

- Explain how the skin functions.

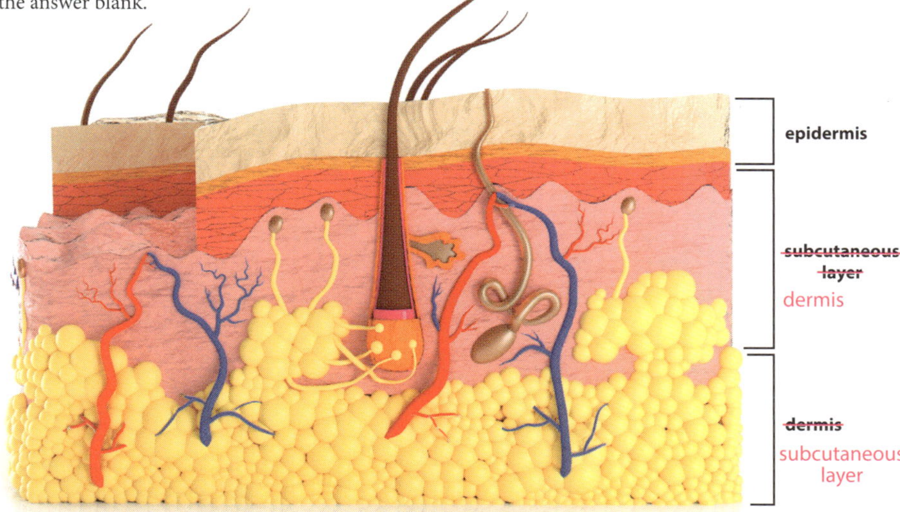

____√____ **1.** Hair, skin, and nails together make up the pulmonary system. (*False*)

________ **2.** The surface of the ~~dermis~~ is made of dead epithelial cells filled with keratin. (*True*)

epidermis

________ **3.** The ~~subcutaneous layer~~ contains oil and wax glands. (*True*)

dermis

____√____ **4.** Fat cells located in the subcutaneous layer help to cushion and insulate the body. (*True*)

____√____ **5.** Hair, which is made of cells filled with oil and wax, grows from hair follicles. (*False*)

________ **6.** Oils released by ~~sweat glands~~ help to keep skin soft and flexible. (*True*)

pores

NOTES

REVIEW 13C

BONES

NAME:

DATE:

Identify the bones numbered in the diagrams at right and below.

1. skull
2. clavicle
3. sternum
4. ulna
5. radius
6. carpals
7. metacarpals
8. pelvis
9. metatarsals
10. tarsals
11. phalanges
12. tibia
13. fibula
14. patella
15. femur
16. phalanges
17. vertebra
18. humerus
19. rib
20. scapula

REVIEW 13C OBJECTIVE

- Identify the major bones of the human body.

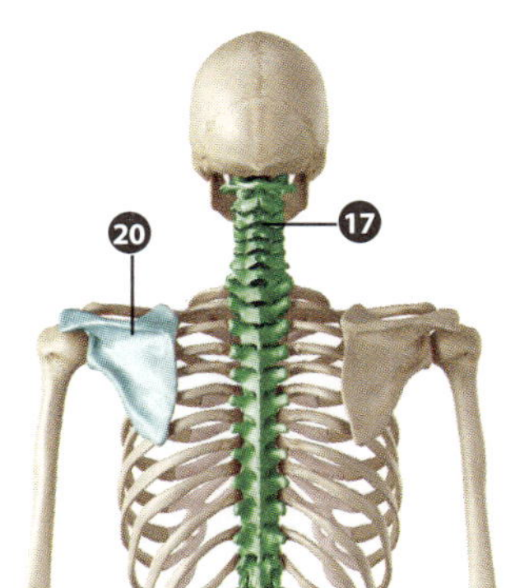

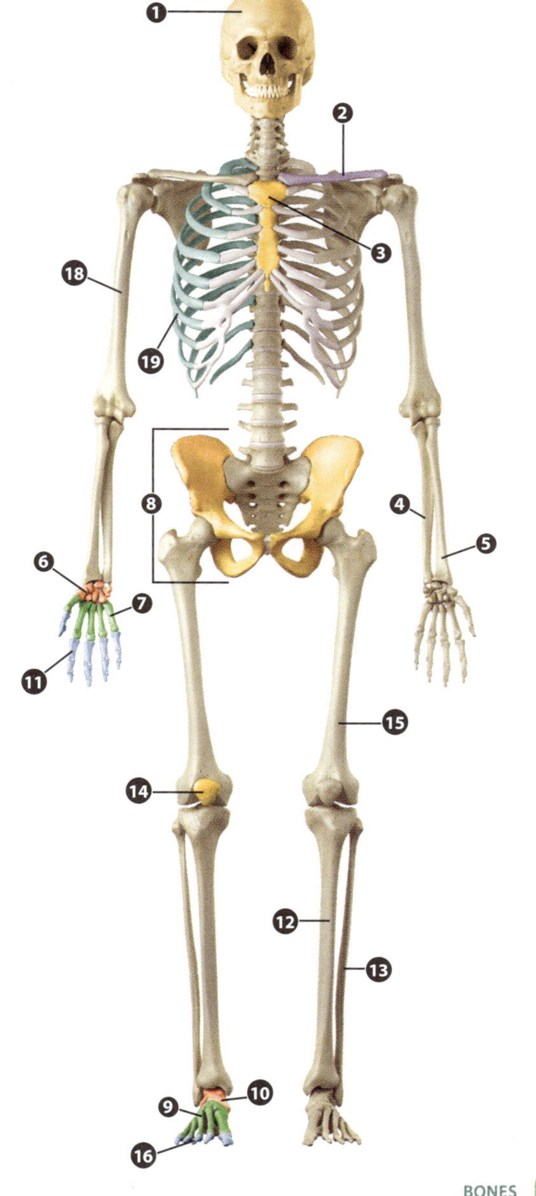

BONES **155**

NOTES

MUSCLES

NAME:

DATE:

REVIEW 13D OBJECTIVE

- Identify examples of voluntary and involuntary muscle.

For each question below, find the muscle on the pictures with the number of the question. Write the name of the muscle on the blank.

biceps	gluteus maximus	pectoralis major	tibialis anterior
deltoid	latissimus dorsi	quadriceps femoris	trapezius
gastrocnemius	oblique	rectus abdominis	triceps

1. gastrocnemius
2. deltoid
3. oblique
4. tibialis anterior
5. triceps
6. rectus abdominis
7. quadriceps femoris
8. biceps
9. trapezius
10. pectoralis major
11. gluteus maximus
12. latissimus dorsi

NOTES

LAB 13E

SKIN DEEP

The Structure of Skin

WHAT STRUCTURES DO WE SEE IN HUMAN SKIN?

Your skin is a major part of your integumentary system. It's easy to take it for granted—until you get a sunburn during summer vacation! After all, your skin is with you every day! But skin is actually quite complex, and it has many important jobs to do. In this lab activity, you'll observe human skin under a microscope to learn more about its structures.

Procedure

Obtain and set up your microscope. Focus your microscope on the slide of human skin. In your field of view, orient the slide so that the epidermis is toward the top and the subcutaneous layer is toward the bottom.

1 Find and observe the epidermis.

1. Where is the epidermis?

It is the top, or outermost, layer.

2. Describe the epidermis.

Answers will vary. It is a multiple-layered tissue of variable

thickness.

3. What type of tissue is the epidermis made of—epithelial, muscular, connective, or nervous?

epithelial

4. What substance fills the epidermal cells before they die?

keratin

NAME:

DATE:

Key Questions 

- What do skin structures look like under a microscope?
- How can I identify the structures of skin?

Equipment 

microscope

prepared slide of a cross section of human skin

- Identify the layers and structures found in human skin.

- Describe the layers and structures found in human skin.

SAVE TIME 

If time is in short supply, set up the microscopes before class.

SKIN DEEP **159**

NOTES

Not all of the structures that students are asked to find for this lab activity are labeled in the Student Edition. Some will be easy to identify even without such labels, such as hair and hair follicles. Others are less obvious. You may need to refer students to additional resources (e.g., online), or you may choose to strike some of these items from the list of things to look for.

Question 10 Answer

Students may not find a sweat gland. Normally it will be in the subcutaneous layer, with the duct passing through the dermis to the surface of the epidermis. Depending on the angle at which the specimen was cut, you may see only a portion of a sweat gland.

---

2. Find and observe the dermis.

5. Where is the dermis?

   It is under the epidermis.

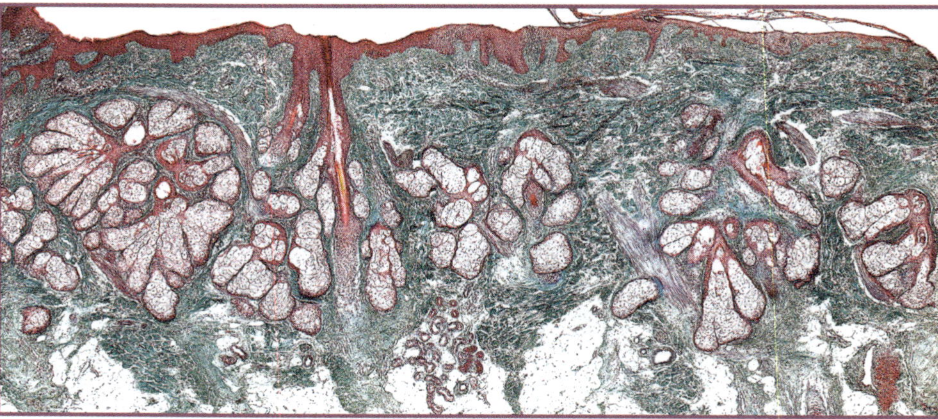

In Questions 6–12, look for the given structure in the dermis. If you can find it, describe its appearance. Leave the line blank for any you cannot find.

6. hair

   Students should find and describe hair. On a preserved slide, they may be able to see only the hair follicle.

7. hair follicle

   Students should find and describe a hair follicle.

8. blood vessel

   Students may not be able to locate blood vessels on a preserved slide.

9. oil gland

   Students should find and describe an oil gland. They are located around hair follicles.

10. sweat gland

    See TE margin for answer.

11. nerve ending

    A nerve ending will be difficult to find even with the best slides.

12. muscle

    Muscle tissue is easy to find on some slides and may be impossible to locate on others.

13. Where are the oil glands located in relation to the hair follicles? What do you think is the significance of this relationship?

Oil glands are located around hair follicles and open into the

hair follicles. The oil glands supply oil to keep the hair and

epidermis soft.

14. Describe the shape of a hair inside the follicle.

Answers will vary. The hair is thicker in the lower part of the hair

follicle. The end of it wraps around blood vessels.

15. Pull a hair out of your scalp. Does the end of the hair from your scalp resemble the end of the hair that you observed through the microscope? If not, how does it differ?

It should appear about the same. The "root" is actually what was

wrapped around the blood vessels.

16. What type of tissue is the dermis mostly made of?

connective tissue

❸ Find and observe the subcutaneous layer.

17. Where is the subcutaneous layer located?

below the dermis

18. Describe the subcutaneous layer in comparison to the dermis and epidermis.

The subcutaneous layer appears less structured and less "full"

than the dermis and epidermis.

19. What are the two main structures that make up the subcutaneous layer?

fibers and fat

20. What structures can you find in the subcutaneous layer?

Student answers will depend on your slides.

Blood vessels and muscles are common. Most of

the subcutaneous layer is fat, which will appear

as open, empty areas.

161

LAB 13F

FIRING UP

Heat from Muscles

DO MUSCLES HAVE OTHER JOBS BESIDES HELPING US TO MOVE?

Humans are endotherms—including you! You've probably heard that the average internal temperature of a human is 37 °C. But how do our bodies maintain this temperature? Muscles use energy to contract—perhaps some of the energy is converted to heat? If so, any additional heat output should be detectable as an increase in temperature.

Procedure

1. What do you think? When muscles are exercised, is some of the energy that they use converted to heat energy? Write a hypothesis that predicts the relationship between exercise and muscle temperature.

    Answers will vary. Example: As muscle use increases, muscle

    temperature will increase.

❶ Choose a person from your lab team to sit where everyone can see him. He should be wearing a short-sleeved shirt (or else roll up his long sleeves).

❷ Put one strip thermometer on each arm, over his biceps. Choose one arm to be the exercised arm. Allow about 15 seconds for the thermometers to indicate the correct temperature for each arm. Record these temperatures in the *Initial* row of Table 1.

❸ With the exercised arm, the student should lift and lower the dumbbell at a constant rate of about once every second. The other arm should remain motionless.

❹ After 30 seconds, take a temperature reading of each arm. It is not necessary to stop the lifting and lowering to record the temperature. Record your data in the appropriate row and column.

NAME:

DATE:

Key Questions 

- How do our bodies maintain their internal temperature?
- Do muscle contractions produce heat?

Equipment

strip thermometers (2)

dumbbell or similar weight

stopwatch or watch with a second hand

LAB 13F OBJECTIVES

- Identify a possible heat source for maintaining homeostasis.
- Compare the heat output of exercised and unexercised muscles.

WEIGHT LIMITS 

Choose a weight that is neither too light (requires little exertion) nor too heavy (causes the student to tire quickly). Depending on the student, a weight in the range of 3–8 lbs should be sufficient. In our tests with a middle school student using a 4 lb weight, 30 repetitions within 30 seconds had little effect on the arm-surface temperature. An additional 30 repetitions made the student strain and raised the arm-surface temperature about 2 °F. The temperature of the unexercised arm was unchanged.

HOMESCHOOL TIP

For a single student, the student should record data while a family member lifts the weight.

NOTES

GRAPHING 

Consider having your students graph their time versus temperature data. A graph for each trial is recommended. They may graph the values for both arms on the same graph.

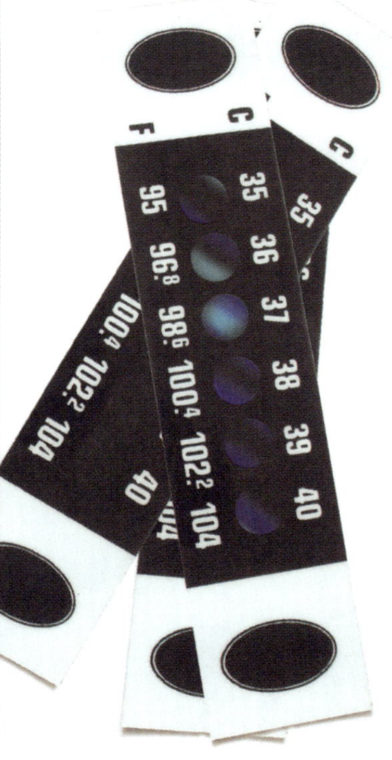

Table 1

Time (sec)	Temperature of exercised arm	Temperature of unexercised arm
Initial		
30		
60		
90		
120		
150		
180		
210		
240		

Table 2

Time (sec)	Temperature of exercised arm	Temperature of unexercised arm
Initial		
30		
60		
90		
120		
150		
180		
210		
240		

Table 3

Time (sec)	Temperature of exercised arm	Temperature of unexercised arm
Initial		
30		
60		
90		
120		
150		
180		
210		
240		

Analysis

2. According to your data, did the temperature of the exercised arm increase, decrease, or remain the same?

It increased.

3. Did the temperature of the unexercised arm increase, decrease, or remain the same?

It remained the same.

4. Do these results support your hypothesis? Explain.

Answers will vary. Example: Yes. As the muscles in the exercised arm contracted, they produced heat, raising the temperature of the exercised arm.

Going Further

5. On the basis of what you learned in this investigation, how do you think that shivering helps your body maintain homeostasis?

Answers will vary. Shivering occurs when the body's temperature drops. The involuntary muscle contractions produce heat energy to help offset the heat energy that is lost to the cold surroundings.

METABOLIC RATE

NAME:

DATE:

Our bodies use a certain number of Calories from the food we eat. Any extra Calories are stored as fat to be used later. Even just lying in bed requires some Calories. This minimum amount of Calories that a person uses per day is called a person's *basal metabolic rate*, or *BMR*. Any activity that a person engages in uses Calories over and above his BMR.

1. Suppose that a person's BMR requires 1500 Calories per day. He exercises every day, causing him to use an additional 1800 Calories per day. If he consumes 4000 Calories per day, what will probably happen?

   Because he is taking in more Calories than his body uses, he will probably store the extra energy as fat and will gain weight.

2. If the same person consumes only 3000 Calories per day, what will probably happen?

   Because he is using more Calories than he is taking in, he will probably use stored energy (fat) that he already has and will lose weight.

3. If an overweight person consumes fewer Calories than his metabolism uses, and he reaches a healthy weight, what might happen if he continues to consume fewer Calories?

   He will eventually use up his fat reserves as he loses too much weight and may even begin to starve.

4. What are some of the nutrients that you need for good health that do not supply energy for metabolism?

   Water, vitamins, and minerals do not supply energy but are essential for proper health.

5. A man has a BMR of 1500 Calories per day and his frequent Ultimate playing causes him to burn an additional 1000 Calories per day. If his breakfast contains 600 Calories and his lunch contained 850 Calories, how many Calories should his supper contain for him to maintain his current weight?

   (1500 Cal + 1000 Cal) − (600 Cal + 850 Cal) = 1050 Cal

REVIEW 14A OBJECTIVE

- Design a plan for sound nutrition incorporating biblical teaching. **BWS**

NOTES

6. Two people eat the same number of Calories each day, but one exercises a great deal while the other exercises very little. Which one is at a greater risk of storing too many Calories? Explain.

The person who does not exercise has the greater risk of storing too many Calories because his caloric requirement is more likely to be lower than his caloric intake.

7. How is monitoring caloric intake and activity being a good steward of the body God has given us?

Monitoring caloric intake and activity helps maintain the health of the body God has given us. This is good stewardship since it also enables a person to more effectively serve God.

REVIEW 14B

THE DIGESTIVE SYSTEM

NAME: ___________________

DATE: ___________________

Your body needs nutrients from food. But you couldn't extract those nutrients without your digestive system. For each question below, write the name of the indicated organ of the digestive system.

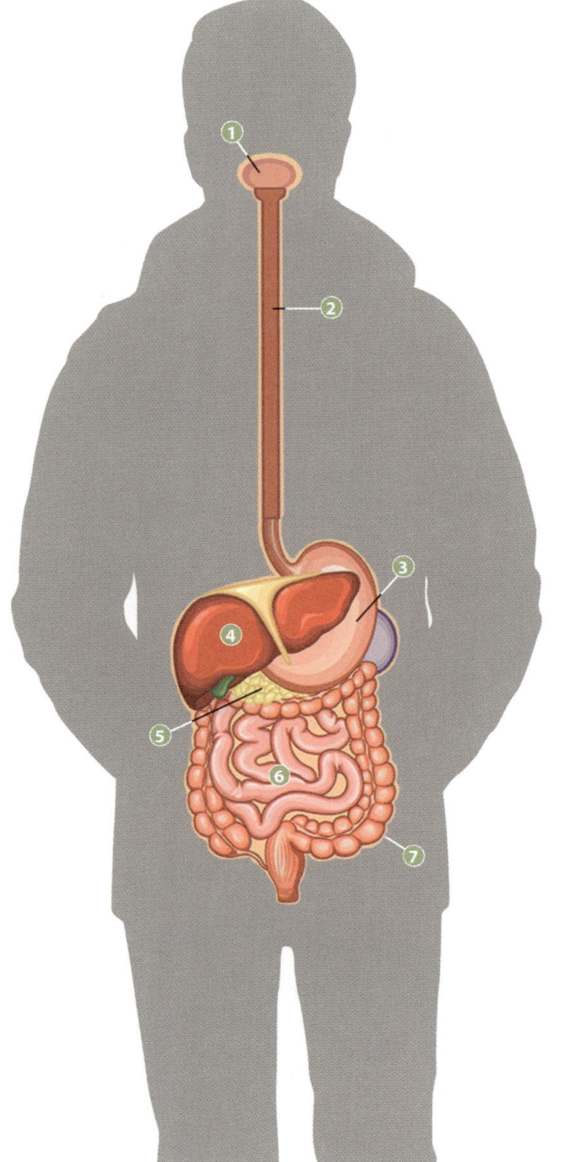

REVIEW 14B OBJECTIVES

- Describe the organs of the digestive system.

- Trace the path of food through the digestive system.

1. mouth ___________________
2. esophagus ___________________
3. stomach ___________________
4. liver ___________________
5. pancreas ___________________
6. small intestine ___________________
7. large intestine ___________________

THE DIGESTIVE SYSTEM **169**

NOTES

____

____

____

____

____

____

THE URINARY SYSTEM

NAME:

DATE:

We all need proteins to live. But proteins come with a potentially deadly price—digesting them produces wastes that contain toxic nitrogen compounds. But don't worry! The urinary system removes such wastes before they can reach dangerous levels.

In the activity below, write the name of the indicated part of the urinary system.

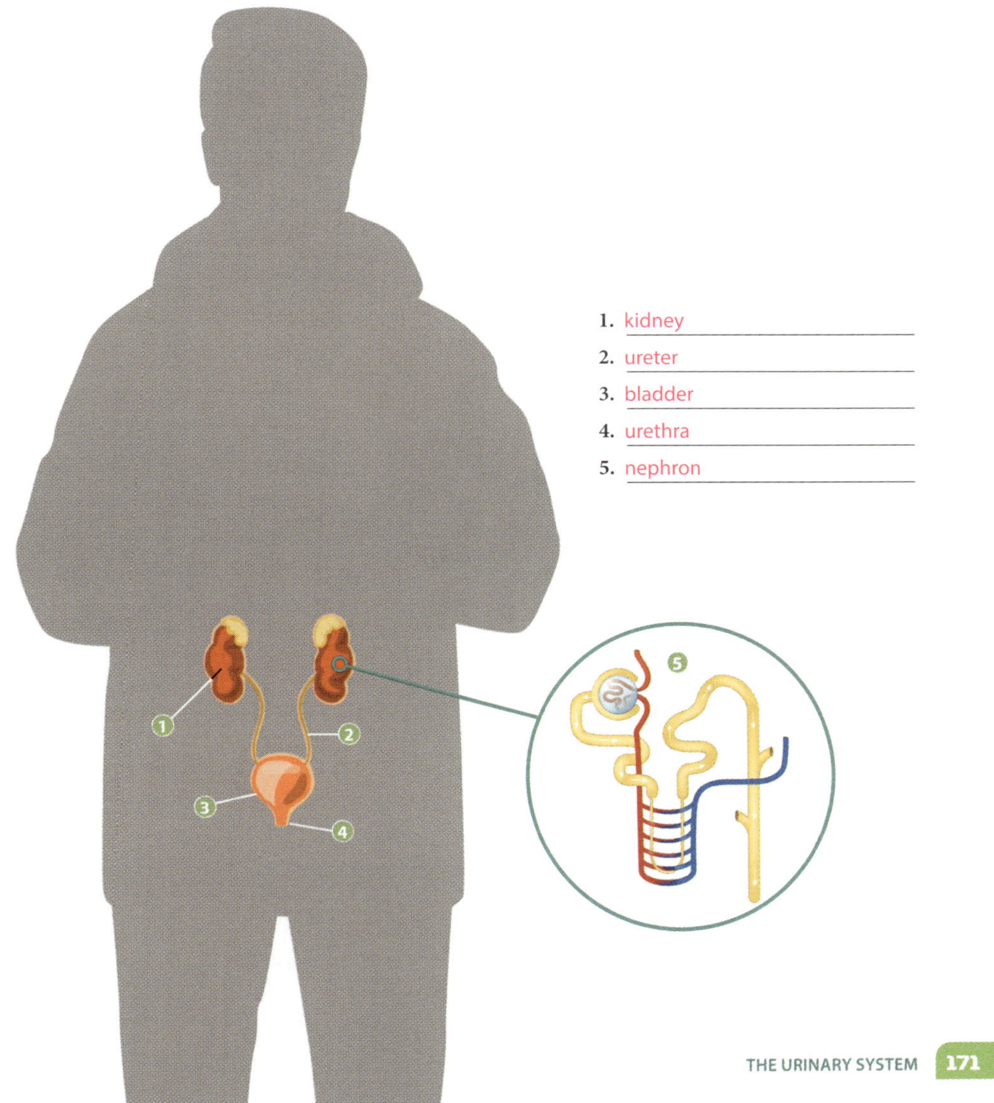

1. kidney
2. ureter
3. bladder
4. urethra
5. nephron

REVIEW 14C OBJECTIVES

- Describe the organs of the urinary system.

- Summarize the way that the kidneys remove waste materials from the blood.

- Trace the path of urine through the urinary system.

NOTES

LAB 14D

USING ENERGY

Burning Calories with Exercise

HOW MANY CALORIES DO I BURN WHEN I EXERCISE?

Many people want to lose weight. Some try exercise, while others try dieting. Some try a combination of the two. How many Calories can you burn while exercising? In this activity you will have a chance to find out. But keep in mind that each person is different, so the results you obtain may be different from the results obtained by your classmate.

Procedure

1. Run for 10 minutes using the running app or pedometer to measure the distance you traveled.

2. Enter the distance and time you ran on Lines 1 and 2 of Table 1.

3. Use an online Calories calculator to calculate the number of calories that you burned during your run. Record that number on Line 3 of Table 1.

4. Use an online Basal Metabolic Rate calculator to determine your BMR. Record your BMR on Line 4 of Table 1.

5. Divide your BMR by 1440 (24 hr × 60 min/hr) to get your average BMR per minute. Multiply that value by the time you ran to determine the total Calories burned due to your BMR during your exercise time. Record that number on Line 5 of Table 1.

6. Subtract the Calories burned due to BMR from the Calories burned during your run to obtain the number of Calories burned due to exercise alone. Record that number on Line 6 of Table 1.

7. Divide the value you calculated in Step 6 by the time you spent exercising in Step 2 to find the number of Calories burned per minute of exercise. Record that number on Line 7 of Table 1.

NAME:

DATE:

Key Questions 

- How can you calculate the number of Calories burned during exercise?
- Is there a connection between exercise and weight?

Equipment

running app or pedometer
stopwatch
online Calories calculator
online BMR calculator

Table 1

	Exercise Data
1. Distance (m)	2000
2. Time (min)	10
3. Calories burned during run	186
4. BMR (Calories/day)	1794
5. Calories burned due to BMR	12.5
6. Calories burned due to exercise alone	173.5
7. Calories burned per minute of exercise	17.4

LAB 14D OBJECTIVES

- Calculate the approximate number of Calories burned during exercise.

- Evaluate the relationship between exercise and weight.

PEDOMETER 

Students with smartphones can easily download a running app. Many are free or have a free version that can provide the measurement needed. Students who use running apps will probably not need a stopwatch, though that may depend on the app used.

Students without smartphones can use a pedometer. They are commonly available; keep in mind that accuracy varies from one to another.

If a running track with a known distance is available, you can have each student run a specified distance while measuring the time he takes to complete the distance. You could also measure and mark a course and have students run that distance while timing themselves.

CALORIES AND BMR 

There are many online Calorie calculators available. The best ones incorporate an individual's weight, height, age, distance run, and duration of a run.

You can also find many online BMR calculators. While they may give somewhat different estimates, they are usually accurate to within a few Calories/day.

SAMPLE DATA 

The data in Table 1 is sample data. Your students' data may be different.

NOTES

1. A small piece of cake with a scoop of ice cream contains about 379 Calories. If you maintained the same pace that you did while running, how long would you have to run to burn this number of Calories?

Answers will vary. The student should divide 379 Calories by the number in Line 7 of Table 1.

Example: 21.8 min

2. If you maintained the pace that you calculated in Step 7 for half an hour, how many Calories would you burn?

Answers will vary. The student should multiply the number of Calories burned per minute by 30 minutes.

Example: 522 Cal

3. On the basis of your results in this activity, do you burn more or fewer Calories than you might have expected for the amount of effort you expend during exercise?

Answers will vary, but students are likely to be surprised by how much effort is required to burn Calories, leading logically to Question 4.

4. What does this suggest to you about using exercise to help maintain your weight?

Answers will vary. Students should mention that exercise burns Calories from the body but exercise alone will probably not burn enough Calories to lose a lot of weight without also controlling one's diet.

Going Further

5. How do nutrition and exercise work in a weight loss program?

Answers will vary. Students should mention that reducing the number of Calories taken in through proper nutrition and increasing the number of Calories used through exercise will give the best results in a weight loss program.

BREAK IT DOWN

Mechanical and Chemical Digestion

HOW DOES CHEWING FOOD HELP DIGEST IT?

Do you like crackers? How about crackers with slices of cheese on them? Crackers are primarily made of starch, a complex carbohydrate. Did you know that your body has a hard time absorbing starch? The molecules are just too big. Saliva contains a digestive enzyme—amylase—that breaks starch down into smaller molecules of glucose.

How does chewing a cracker fit into this? In this activity you will investigate whether chewing affects how well the starch in a cracker is digested.

Modeling Digestion

1. Use a marking pencil to label five test tubes *A* through *E*.

2. In Tube A, place a crushed piece of cracker. Add just enough water to moisten the cracker.

3. Add 5 mL of amylase solution to Tube B.

4. Place a cracker in a bowl and pour 5 mL of amylase solution on it. After 5 seconds, pour the amylase off. Break the cracker into smaller pieces and place them in Tube C.

5. Place a cracker in the mortar and pour 5 mL of amylase solution on it. Without grinding it, firmly tamp it twice with the pestle. Pour off the amylase solution, break up the cracker (if needed), and place it in Tube D.

6. Place another cracker in the mortar and pour 5 mL of amylase on it. Grind it with the pestle until it is a soft mass. Pour off any excess amylase solution and place the mass in Tube E.

NAME:

DATE:

Key Question

- How do mechanical and chemical digestion work together?

Equipment

test tubes (5)
bowl
mortar and pestle
marking pencil
saltine crackers
water
graduated cylinder, 10 mL
hot plate
beaker, 600 mL
test tube rack
test tube tongs
Benedict's solution
amylase solution
goggles

LAB 14E OBJECTIVE

- Distinguish between the effects of mechanical and chemical digestion.

BENEDICT'S SOLUTION 

Benedict's solution is an eye irritant, so eye protection is very important.

Prior to attempting this lab, research the disposal requirements for Benedict's solution.

NOTES

BENEDICT'S SOLUTION RESULTS ✓

A student with color vision deficiency may struggle to distinguish between the colors that result from Benedict's solution. For that reason, any such students should be paired with a student having normal color vision.

Question 7 Answer

Answers will vary. Students' responses should be consistent with their hypotheses.

Analyzing Benedict's Solution in the Test Tubes

If no glucose is present in a substance, it will remain blue.

If a small amount of glucose is present, it will turn yellow.

If a medium amount of glucose is present, it will turn orange.

If a large amount of glucose is present, it will turn brick red.

1. What does the amylase solution in this lab represent?

   saliva

2. What does the use of the mortar and pestle represent?

   the grinding action of chewing

3. Which tube do you think will contain the most digested cracker? Write your answer as a hypothesis.

   Answers will vary. Example: Tube 5 will contain the most

   digested cracker.

Testing for Starch

1. Add 5 mL of Benedict's solution to each test tube. When heated, Benedict's solution will change color if glucose is present.

4. What is the purpose of Tube A?

   Tube A functions as a control to show the amount of glucose

   in an undigested cracker.

5. What is the purpose of Tube B?

   Tube B ensures that the amylase solution does not contain

   glucose.

2. Place the test tubes in a large beaker containing 400 mL of water.

3. Heat the beaker on a hot plate to boiling.

4. Remove the test tubes to a test tube rack to cool.

5. Record the color of the five test tubes in the Color column of Table 1.

6. Using the information in the margin at left, complete the *Amount of glucose present* column of Table 1.

6. Which test tube had the most digested cracker?

   Tube E

7. Was your hypothesis correct?

   See TE margin for answer.

8. Explain your results using what you have learned about mechanical and chemical digestion.

   The enzyme amylase transforms starch into glucose. Grinding

   the cracker with the amylase gives the enzyme more surface

   area to work on and mixes the amylase more thoroughly with

   the cracker.

9. On the basis of your results, which method of eating a cracker results in greater digestion: swallowing it whole, chewing it 5 times, or chewing it 15 times? Explain.

Chewing it 15 times would result in more thorough digestion.

More chewing breaks the cracker into smaller pieces and mixes

the saliva and cracker more thoroughly, allowing the amylase to

digest more starch.

10. Amylase ceases to function at a low pH due to the presence of a strong acid. Why would starch digestion stop when food reaches the stomach?

The low pH in the stomach caused by stomach acid causes

the amylase from saliva to cease to function.

Table 1

Tube	Color	Amount of glucose present
A	blue	none
B	blue	none
C	yellow	small amount
D	yellow/orange	medium amount
E	orange/red	large amount

INTESTINAL AMYLASE

While the stomach acid denatures the amylase from saliva, additional amylase from the pancreas is added to food in the small intestine. Since the small intestine neutralizes the acidic chyme coming from the stomach, the amylase is free to act on the remaining starch.

THE RESPIRATORY SYSTEM

NAME:

DATE:

Do you enjoy any kind of sports? If so, then you know that many sports make you huff and puff as your muscles use lots of glucose and oxygen.

For Questions 1–6, write the name of the indicated part of the respiratory system shown below in the diagrams using the structures in the list below.

alveolus	capillary	pharynx
bronchus	lung	trachea

1. pharynx
2. trachea
3. bronchus
4. lung
5. alveolus
6. capillary

7. According to the diagram below, which blood vessels contain deoxygenated blood? Explain.

   The pulmonary arteries contain

   deoxygenated blood because they are

   carrying it from the heart to be

   oxygenated by the air in the lungs.

REVIEW 15A OBJECTIVES

- Describe the organs of the respiratory system.

- Trace the flow of air through the respiratory system.

NOTES

THE HEART

NAME:

DATE:

Your heart is the only organ in your body that has cardiac muscle. In fact, it is one large muscle. It works all the time, no matter what you are doing, and your life depends on it.

Each of the phrases in Questions 1–7 below describes a part of the heart. Use the number of the description to label it in the diagram below. Then write the name of the part in the blank.

- Describe the components of the circulatory system.

- Trace the flow of oxygen and carbon dioxide through the body.

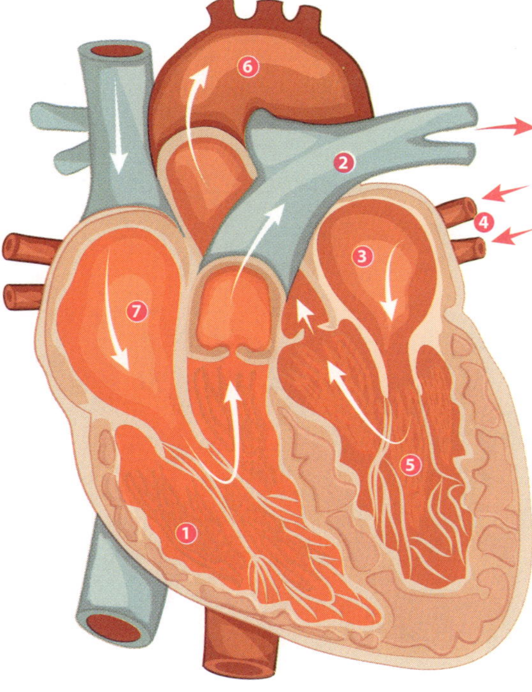

1. pumps blood to the lungs

   right ventricle

2. carries deoxygenated blood to the lungs

   pulmonary artery

3. receives oxygenated blood from the lungs

   left atrium

4. returns oxygenated blood to the heart

   pulmonary vein

5. pumps blood into the aorta

   left ventricle

6. carries oxygenated blood from the heart to all parts of the body

   aorta

7. receives blood from the body

   right atrium

8. Draw arrows on the diagram to show the flow of blood through the heart and lungs. Begin at the right atrium and end at the aorta.

   See diagram.

NOTES

REVIEW 15C

THE LYMPHATIC SYSTEM

NAME:

DATE:

Your lymphatic system is sort of like the utility pipes under the streets of a city. It's hidden and you don't think about it a lot, but your body wouldn't be able to function without it.

Using the image below, give the name and function of each indicated structure.

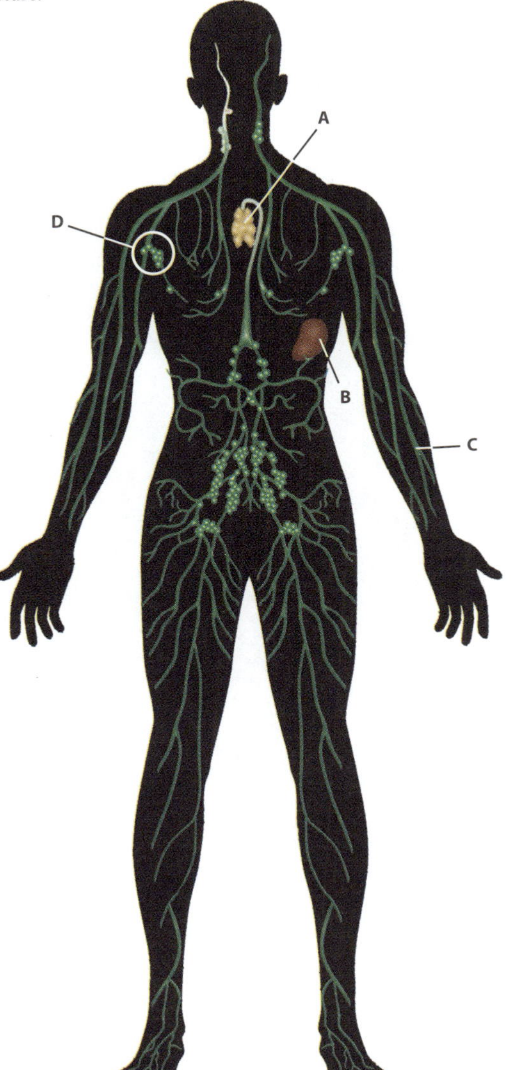

1. Structure A

   thymus; White blood cells mature in the thymus, though its size and activity greatly decrease by adolescence.

2. Structure B

   spleen; The spleen recycles worn-out red blood cells and produces and stores some white blood cells.

3. Structure C

   lymph vessel; Lymph vessels conduct plasma that leaks from capillaries back to the circulatory system.

4. Structure D

   lymph node; Lymph nodes filter dead cells, bacteria, viruses, and other particles out of the lymph.

REVIEW 15C OBJECTIVES

- Describe the components of the lymphatic system.

- Explain how the lymphatic system works together with other systems in the body.

NOTES

LAB 15D

I'LL HUFF, AND I'LL PUFF

Increasing Respiration Rate

NAME:

DATE:

DOES EXERCISE AFFECT MY BREATHING RATE?

As you learned in your textbook, your lungs supply your body with oxygen and remove carbon dioxide. The oxygen is carried by your blood to all the cells of your body. If your muscles are doing a lot of work, they use more oxygen for cellular respiration than they typically do. So what happens when your body needs more oxygen? Do you breathe faster?

In this activity, you will test that question by measuring your respiration rate—the number of breaths you take in a minute.

Key Questions

- Does respiration rate increase as a result of activity level?
- How long does it take the respiration rate to return to normal after exercise?

Equipment

respiration rate app or stopwatch

Procedure

1. Do you think that exercise will increase your respiration rate? Write your answer as a hypothesis.

   Answers will vary.

   Example: Exercise will increase my respiration rate.

For Steps ❶ – ❻, team up with a classmate so that you can count each other's breaths.

❶ Sit quietly for three minutes. Then have your partner count the number of breaths you take as you breathe normally for one minute. Record your respiration rate in the appropriate row of Table 1.

2. Why do you take your first respiration rate reading after sitting quietly for several minutes?

   Answers will vary. Sitting quietly before taking the first

   respiration rate reading allows the student to establish a baseline

   for his respiration rate.

LAB 15D OBJECTIVES

- Determine whether respiration rate increases with amount of exercise.

- Measure the time it takes for respiration rate to return to normal after exercise.

CROSS CURRICULUM 

If you coordinate with the physical education teacher, students could collect data in gym class and then use science class time to analyze the data.

PREPARATION 

This lab activity requires a route for your students to walk and run. A running track would work well. The straightaways of many tracks are 100 m long. If you don't have access to a track, outline a route approximately 200 m long. The exact length is not crucial.

MEASURING RESPIRATION RATE 

Measuring respiration rate is not difficult. Since respiration can be controlled voluntarily, it is best to have another person count the number of breaths that a person takes in a minute. The counter simply counts the number of times the other person's chest rises in a minute. There are some free apps that may speed up the process, but not having access to them will not greatly affect this lab activity.

NOTES

Table 1

Activity	Respiration Rate (breaths/min)
Sitting	16
Walking slowly	18
Walking briskly	20
Jogging	28
Running	40

2. Slowly walk the route outlined by your teacher. Then measure and record your respiration rate in the appropriate row of Table 1.

3. Briskly walk the same route. Measure and record your respiration rate in the appropriate row of Table 1.

4. Jog the same route. Measure and record your respiration rate in the appropriate row of Table 1.

5. Run the same route. Measure and record your respiration rate in the appropriate row of Table 1.

6. As soon as you finish Step 5, start your stopwatch and sit quietly, periodically checking your respiration rate until it returns to the same rate that you measured in Step 1.

3. How long did it take for your respiration rate to return to normal?

Answers will vary. Example: 16 min

Save your answer to Question 3. You will need it to complete the *Going Further* section in Lab 15E.

4. Did your respiration rate change as a result of your exercise?

Yes.

5. Was your hypothesis correct?

Answers will vary. Make sure that each student's answer is

consistent with his original hypothesis.

Going Further

6. What would happen if your breathing remained unchanged during exercise?

Muscles would not be able to maintain cellular respiration

as long.

LAB 15E

THE DRUMMING OF THE HEART

Increasing Heart Rate

DOES EXERCISE AFFECT MY HEART RATE?

You know that your heart pumps blood through the circulatory system, carrying oxygen and glucose to your cells. But what happens when your cells need more oxygen and glucose, like when you are exercising? Does the heart pump faster?

In this activity, you are going to try to answer that question.

Procedure

1. Do you think that exercise will increase your heart rate? Write your answer as a hypothesis.

   Answers will vary.

   Example: Exercise will not affect my heart rate.

❶ Sit quietly for three minutes. Then measure and record your pulse (which is also your heart rate) in the appropriate row of Table 1.

2. Why do you take your first pulse reading after sitting quietly for several minutes?

   Answers will vary. Sitting quietly before taking the first pulse

   reading allows the student to establish a baseline for his heart

   rate.

NAME:

DATE:

Key Questions

- Does heart rate increase as a result of activity level?
- How long does it take the heart rate to return to normal after exercise?

Equipment

pulse monitor or pulse monitor app

stopwatch

LAB 15E OBJECTIVES

- Determine whether heart rate increases with amount of exercise.

- Measure the time it takes for the heart rate to return to normal after exercise.

PREPARATION 

See Preparation teacher note on page 185.

If you assigned Lab 15D and intend to have your students complete the *Going Further* section for this activity, the same route should be used for both labs.

PULSE MONITOR 

Pulse monitors are not expensive, but supplying one to each student may be cost prohibitive. Students with smartphones can download pulse monitors. You may need to instruct some students without smartphones how to measure their pulses.

THE DRUMMING OF THE HEART 187

NOTES

SAMPLE DATA 

The data in Table 1 is sample data. Your students' data may be different.

KEEP HYDRATED

During Step 6, make sure that students do not interpret sitting quietly as precluding them from drinking liquids if they need to. Keeping hydrated is important.

CONNECTION WITH PREVIOUS LAB

The *Going Further* section connects this lab with Lab 15D. If you did not assign Lab 15D, this *Going Further* section cannot be completed.

Table 1

Activity	Heart Rate (beats/min)
Sitting	82
Walking slowly	92
Walking briskly	96
Jogging	152
Running	156

Table 2

	Heart Rate	Respiration Rate
Time to return to sitting rate (min)	16	16

2 Slowly walk the route outlined by your teacher. Then measure and record your heart rate in the appropriate row of Table 1.

3 Briskly walk the same route. Measure and record your heart rate in the appropriate row of Table 1.

4 Jog the same route. Measure and record your heart rate in the appropriate row of Table 1.

5 Run the same route. Measure and record your heart rate in the appropriate row of Table 1.

6 As soon as you finish Step **5**, start your stopwatch and sit quietly, periodically checking your pulse. In the *Heart Rate* column of Table 2, record the time it takes for your heart rate to return to the same rate that you measured in Step **1**.

3. Rank the five activities on the basis of heart rate, with the slowest first.

sitting, walking slowly, walking briskly, jogging, running

4. Did exercise affect your heart rate?

Yes.

5. Was your hypothesis correct? Explain.

Answers will vary. Make sure that each student's answer is consistent with his original hypothesis.

6. If there was a change in your heart rate as a result of exercise, what type of cells likely uses much of the extra oxygen and glucose being carried by the blood?

muscle cells

Going Further

1 Retrieve from Lab 15D the time that it took for your respiration rate to return to normal. Copy that time into the *Respiration Rate* column of Table 2.

7. Were the times similar?

Answers will vary. They will likely be similar.

8. On the basis of what you have learned, suggest an explanation for your observation in Question 7.

Answers will vary. Since the respiration rate increases in order to supply the needed oxygen, the heart rate also increases in order to bring oxygen to muscle cells. As the oxygen debt is met, both the heart and respiration rates decrease.

IMMUNE SYSTEM ANALOGY

NAME:

DATE:

REVIEW 16A OBJECTIVE

• Describe the body's lines of defense.

Have you ever been to a castle? Ancient fortresses were originally created to defend the people who lived inside from attack. In addition to the walls, castles also had different types of soldiers who could fight off enemy armies. Your immune system is much like a castle with its defending army. It contains different components that work together to fight off pathogens (the invading army) to protect the castle owner (you!).

Each statement below names and describes one of the defenses of a castle. Write the part of the immune system from the list below that each castle defense represents in this analogy.

antibodies	**inflammation**	**skin**
B-cells	**macrophages**	**T-cells**
fever	**mucous membrane**	

1. The castle walls provide an outer barrier around the castle.

   skin

2. Arrows are shot at enemies.

   antibodies

3. When the archers, men-at-arms, and knights begin defending the castle, all ordinary life becomes much more difficult or even ceases altogether.

   inflammation

4. The portcullis is a metal grate that helps protect the one main entrance to the castle.

   mucous membrane

5. Hot water or oil could be poured down on enemy attackers.

   fever

6. Soldiers use their swords to fight enemies that have come over the wall.

   T-cells

7. The men-at-arms patrol the castle wall during times of peace in case an enemy sneaks over the wall.

   macrophages

8. Archers shoot arrows at attackers.

   B-cells

NOTES

REVIEW 16B

THE NERVOUS SYSTEM

NAME:

DATE:

Tightrope walkers depend on a combination of physics and precise coordination to successfully walk on a thin line. While your coordination may not be as great as a tightrope walker's, every action you take, such as throwing a ball, typing on a keyboard, or just walking across a room, requires great coordination. God has designed our nervous system so marvelously that we usually don't even think about how much coordination we use every day.

In this activity, write the part of the nervous system from the list below described by each statement.

axon	motor neuron
brain	nervous system
cell body	neuron
central nervous system	peripheral nervous system
dendrite	reflex
impulse	sensory neuron
interneuron	spinal cord
motor control	synapse

REVIEW 16B OBJECTIVES

- Describe the parts of the nervous system.

- Summarize how the nervous system relays messages.

- Explain how the parts of the nervous system work together to control the body and maintain homeostasis.

1. The coordination of all the muscles that work together to complete a motion:

   motor control

2. The part of the neuron that contains the nucleus:

   cell body

3. A nervous system cell that is capable of transmitting impulses:

   neuron

4. The part of the central nervous system that is found in the skull:

   brain

5. The system of the body that includes the brain and nerves:

   nervous system

6. An extension of a neuron that carries impulses away from the cell body:

   axon

7. An electrical signal that controls your body:

   impulse

8. The division of the nervous system that is composed of the brain and the spinal cord:

   central nervous system

9. A neuron that is in a reflex arc and serves as a go-between for the other cells in the reflex arc:

   interneuron

10. The space that impulses jump across by means of special chemicals:

    synapse

11. The division of the nervous system that is composed of the nerves outside of the brain and spinal cord:

    peripheral nervous system

12. An immediate, inborn response to a stimulus:

    reflex

13. The part of the central nervous system not found in the skull:

    spinal cord

14. In a reflex arc, the neuron that receives the stimulus and initiates an impulse:

    sensory neuron

15. An extension of a neuron that carries impulses toward the cell body:

    dendrite

16. In a reflex arc, the neuron that carries an impulse to a muscle:

    motor neuron

NOTES

EYES AND EARS

NAME:

DATE:

Your eyes and your ears help you navigate the world around you. Your sight is your dominant sense, but your hearing is also very important.

In Questions 1–6, write the letter used to indicate the part of the eye in the image below. Then write the definition for the part.

REVIEW 16C OBJECTIVE

- Describe the sensory organs of the body.

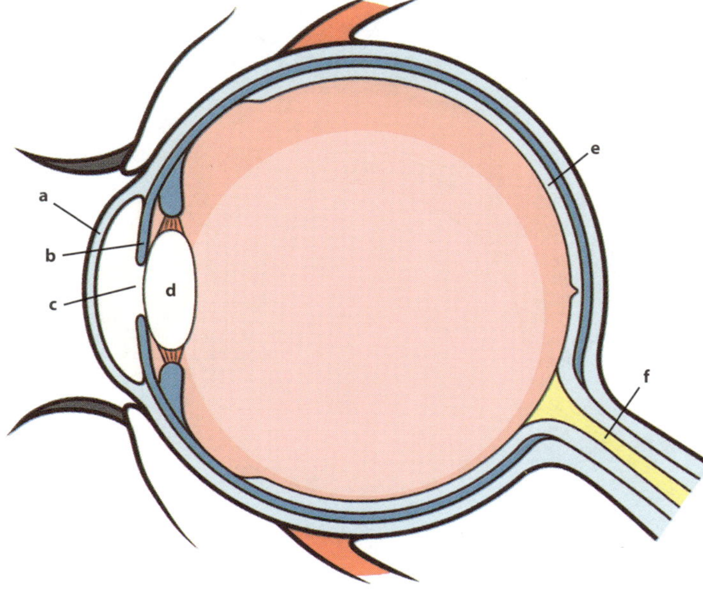

1. retina

e: the inner layer of the eye that contains the light-sensitive receptors

2. optic nerve

f: the nerve that carries impulses from the eye to the brain

3. lens

d: the structure that focuses light waves onto the retina

4. pupil

c: the opening in the iris through which light enters the lens

5. cornea

a: the clear layer through which light enters the eye

6. iris

b: the colored part of the middle layer of the eye that controls the size of the pupil

NOTES

For Questions 7–16, write the letter used to indicate the part of the ear in the image below. Then write the definition for the part.

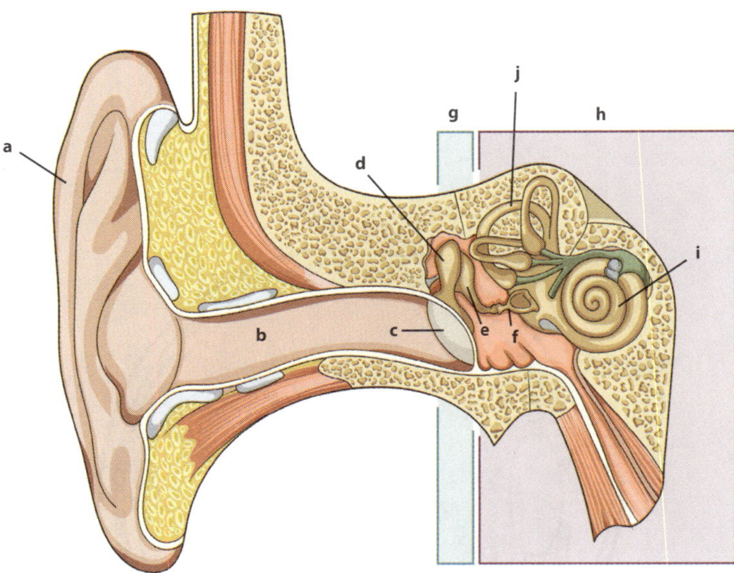

7. ear flap

   a: the structure that collects sound waves

8. hammer

   d: the ear bone that connects the eardrum to the anvil

9. middle ear

   g: the air-filled chamber that contains the ear bones

10. inner ear

    h: the section of the ear that contains the cochlea and the semicircular canals

11. semicircular canal

    j: the structure that senses body balance and position

12. cochlea

    i: the coiled, tubular structure in the inner ear that contains fluid

13. stirrup

    f: the ear bone that connects the anvil and the cochlea

14. anvil

    e: the ear bone that connects the stirrup and the hammer

15. eardrum

    c: the thin membrane that makes up the innermost part of the outer ear

16. ear canal

    b: the tube that permits sound waves to reach the eardrum

LAB 16D

HOT OR COLD?

The Skin's Sensation of Temperature

HOW SMALL OF A DIFFERENCE IN TEMPERATURE CAN YOUR SKIN DETECT?

Have you ever stepped into a pool only to find that the water was freezing? After a few minutes, you might find that the temperature really isn't that bad. Just how accurate are the temperature receptors in your skin? In this activity, you will have a chance to find out.

Procedure

➊ Team up with several classmates to decide which of you will test the water.

1. What do you think is the least temperature difference between two cups of water that a person can distinguish? Write your answer as a hypothesis.

   Answers will vary. Example: A tester can distinguish between

   two temperatures that are at least 3 °C apart.

➋ Use the random number generator to generate a number between 0 and 5, inclusive.

NAME:

DATE:

Key Question

- How accurate are the temperature receptors in your skin?

Equipment

large glasses or cups (2)
cold water
hot water
thermometers (2)
random number generator

LAB 16D OBJECTIVE

- Evaluate the accuracy of the temperature receptors in the skin.

EQUIPMENT 

Large foam cups work best. A large water volume and the insulating nature of foam (polystyrene) will limit temperature changes for the short time they are needed.

If you have a digital thermometer that gives a reading almost instantaneously, this investigation can be done with only one thermometer. But if you have analog thermometers, using only one requires continually moving the thermometer to different glasses and allowing time for readings to stabilize, so it is easier to have a thermometer for each glass. This allows students to more easily maintain the temperatures of each glass.

Random number generators can be easily found online.

HOT OR COLD? **195**

NOTES

❸ Find the two temperatures associated with that number in the *Glass 1* and *Glass 2* columns of Table 1.

❹ Fill two glasses with hot and cold water until the temperatures of the water in Glass 1 and Glass 2 reach the temperatures you identified in Step **❸**. Do not let the tester know whether the temperatures are the same or different.

❺ Have the tester dip two fingers in Glass 1 and two in Glass 2. The tester should then tell you whether he thinks the temperatures of the water in the cups are the same or different. Enter whether he was correct (C) or incorrect (I) in the *Test 1* column of the appropriate row of Table 1.

❻ Repeat Steps **❷** – **❺** until three tests have been run on every row of temperatures in Table 1.

TABLE 1 DATA

Since the students are using a random number generator, it is quite likely that certain rows of Table 1 will be filled while other rows have only one or no cells filled. Assure students that this is perfectly acceptable. If one row is filled and the random number generator returns the number of that row, instruct students to generate another number until all rows have been tested three times.

Table 1

Random Number	Glass 1	Glass 2	Test 1	Test 2	Test 3
0	34 °C	34 °C			
1	34 °C	33 °C			
2	34 °C	32 °C			
3	34 °C	31 °C			
4	34 °C	30 °C			
5	34 °C	29 °C			

2. What was the minimum degrees difference that the tester could correctly distinguish *all three times*?

Answers will vary.

3. At how many degrees difference did the tester *incorrectly* say that the cups were the same all three times?

Answers will vary.

4. Was there a range of difference where the tester could sometimes tell a difference and sometimes not? If so, what was that range?

Answers will vary.

5. Was your hypothesis correct?

Answers will vary.

Going Further

6. On the basis of your experience and the experience described in the introduction, do you think that immersing your hands in ice water for thirty seconds (definitely not recommended!) before doing this experiment would skew the results?

Answers will vary. It probably would change the results.

LAB 16E

GHOST OF COLORS PAST

Afterimages

WHAT CAUSES AFTERIMAGES?

Have you ever looked at a bright light and then looked away only to see a bright light where there was none? This effect is called an *afterimage*. After the cones in your eyes receive bright light, they become less sensitive to light for a few seconds. During that time, those cones do not detect light as well as the cones around them.

Your eyes have three different types of cones. Each type perceives one of three colors: red, green, or blue. If bright light causes one type of the cones in an area of your eye to become less sensitive, the other cones continue to function normally. This causes an afterimage.

In this activity, you will experiment with afterimages.

Procedure

1. What color of afterimage do you think green will cause? How about the other colors? Fill in the *Predicted Afterimage* column of Table 1 with your answers. This will function as your hypothesis. (The colors listed in the table are the same colors shown in Color Circles 1 and 2.)

2. Make sure that you are in a brightly lit area. Place Color Circle 1 about 50 cm away from your face.

3. Stare at the dot in the center of the circle for 45 s. Keep your eyes on the dot for the entire time.

4. After 45 s have passed, immediately look at a piece of white paper, observing what you see (the afterimage).

1. Is the shape of the afterimage the same or different from Color Circle 1?

   It is the same.

2. Are the colors of the afterimage the same or different from Color Circle 1?

   They are different.

5. Rest your eyes for several minutes and then repeat Steps 2–4, noting the colors you see in the afterimage in relationship to the colors in Color Circle 1. Record the colors in the *Actual Afterimage Color* column of Table 1.

6. Repeat Steps 2 – 5 for Color Circle 2.

3. How many colors did you successfully predict?

   Answers will vary.

NAME:

DATE:

Key Question

- How are afterimage colors related to the colors of real images?

Equipment

white paper (2 sheets)
stopwatch

LAB 16E OBJECTIVE

- Evaluate the relationship between the colors of real images and those of afterimages.

COLOR DIFFERENTIATION 

Because this lab requires students to identify colors, it is not recommended for students with color-vision deficiency.

BRIGHTLY LIT ROOM

To obtain a good afterimage, students will need a brightly lit room. They may need a lamp placed near the paper they are staring at. Alternatively, you might display the color circles and negative flag onto a screen or a wall.

NOTES

Table 1

Afterimages

	Original Color	Predicted Afterimage Color	Actual Afterimage Color
Color Circle 1	Green		magenta (pink)
	Blue		yellow
	Red		cyan (blue-green)
Color Circle 2	Magenta		green
	Yellow		blue
	Cyan		red

4. Using the information that you entered in Table 1, why do you think surgeons wear green or blue scrubs when performing operations? (*Hint*: During surgery, surgeons see a lot of red.)

See TE margin for answer.

5. White light is a combination of all colors of light. Why might a room with bright white walls tire your eyes?

It would tire all three types of cones.

Going Further

Shown at the bottom of this page is an image of a national flag with the colors changed.

❶ Research the flags of the countries in the box below.

Chad	France	Ireland	Nigeria

❷ Follow the steps in the *Procedure* section to see the afterimage of the image below.

6. What country does the flag belong to?

Ireland

Color Circle 1

NAME: ______________________

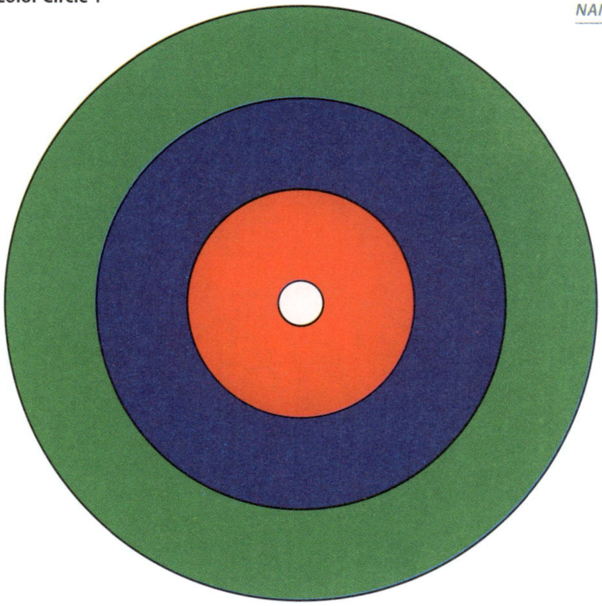

- green
- blue
- red
- magenta
- yellow
- cyan

Color Circle 2

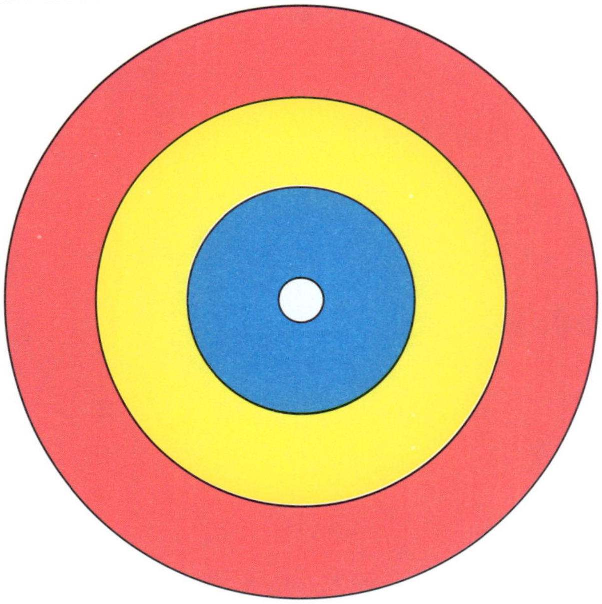

COMPLEMENTARY COLORS

Complementary colors are green and magenta, blue and yellow, and red and cyan. As a result, the afterimage of Color Circle 1 will be Color Circle 2 and vice versa.

GHOST OF COLORS PAST 199

THE ENDOCRINE SYSTEM

NAME:

DATE:

In football, quarterbacks get all the glory. People don't usually talk about the stats of an offensive lineman, but he is vital for winning the game. When it comes to internal controls in the body, the nervous system gets all the attention. After all, it includes the brain! But the endocrine system is also really important for bodily control.

Complete the graphic organizer below, comparing the glands of the endocrine system, by filling in each blank cell with the appropriate information from pages 352–56 of your textbook.

Glands of the Endocrine System

Gland	Location	Hormone Produced	Function
Adrenal gland	on top of the kidneys	epinephrine	prepares the body for emergencies
Hypothalamus	the brain	various	controls the endocrine system
Ovary	female abdomen	estrogen	helps females reach physical maturity
Pancreas	abdomen	insulin, glucagon	regulates glucose in the bloodstream
Pituitary gland	the brain	growth hormone	helps control the body's processes
Testis	male groin	testosterone	helps males reach physical maturity
Thyroid	throat	various	controls how fast the body burns food; affects growth

THE ENDOCRINE SYSTEM 201

REVIEW 17A OBJECTIVES

- Describe the parts of the endocrine system.
- Explain how the structures of the endocrine system function.

NOTES

REVIEW 17B

REPRODUCTION

NAME:

DATE:

- List the structures and functions of the reproductive organs.

The reproductive system is a gift from God that allows us to fulfill the Creation Mandate. At the same time, though, God has laid out the very specific guideline that human sexual activity is strictly for marriage between a man and a woman. And because it's a gift from God, it's important to understand it.

For each of the phrases below, write the structure or process described.

1. the main organs of the reproductive system

   the gonads

2. male gametes

   sperm

3. a fluid that nourishes and protects male gametes

   semen

4. male gonads

   testes

5. female gonads

   ovaries

6. the process of male and female gametes uniting

   fertilization

7. a structure that nourishes and protects a developing baby

   uterus

8. a female gamete

   ovum

9. a one-cell baby formed immediately after gametes unite

   zygote

10. a multicell baby before eight weeks of development

    embryo

11. the process of a baby attaching to the inside of a mother's nourishing structure

    implantation

REPRODUCTION **203**

NOTES

GROWING UP

NAME:

DATE:

Growing up is an exciting and sometimes traumatic time of life, full of excitement, sorrow, and joy. Some of the milestones of growing, both before and after birth, are listed below. Indicate in which stage of growth each milestone occurs, using the stages listed below.

first trimester	**third trimester**	**adolescence**
second trimester	**childhood**	

1. A baby begins to suck his thumb.

   second trimester

2. Communication begins.

   childhood

3. A baby develops from an embryo into a fetus.

   first trimester

4. Hair grows.

   third trimester

5. Arms and legs get longer in proportion to the rest of the body.

   childhood

6. A baby is about the size of a cantaloupe.

   third trimester

7. Puberty occurs.

   adolescence

8. The placenta connects the baby to the mother.

   first trimester

REVIEW 17C OBJECTIVES

- List the stages of human development.

- Distinguish the different stages of human development on the basis of structure and function.

NOTES

TOO MUCH SUGAR

Inquiring into Trends in Diabetes

HOW MANY AMERICANS HAVE DIABETES?

Do you know someone with diabetes? Many Americans live with this disease of the endocrine system. The pancreas produces the hormone insulin that causes cells to take in glucose from the blood to use as energy. But for a diabetic, something goes wrong with this process.

Most diabetics have Type 2 diabetes. Their pancreases still produce insulin, but their cells do not respond correctly. In this activity, you will research how the number of cases of diabetes in the United States has changed over the years. You will also investigate some potential causes for any changes.

Procedure

PLANNING/WRITING GUIDING QUESTIONS

1. Conduct initial research into Type 2 diabetes. This initial research is not meant to be complete but should be sufficient to give you direction in your inquiry. Doing Internet searches on the following topics may be helpful:

   - For how many years are there accurate statistics for the number of Type 2 diabetics in the United States?
   - Has the number of Type 2 diabetics in the United States changed greatly over the past few decades?
   - What are the long-term health consequences of diabetes?
   - What lifestyles are associated with an increased risk of Type 2 diabetes?

2. Identify at least three websites with information on Type 2 diabetes.

3. On the basis of your prior knowledge, the class discussion, and your initial research, brainstorm with your lab team to identify issues related to Type 2 diabetes.

4. Write some specific questions related to these issues that you could answer by additional research.

NAME:

DATE:

Key Questions 

- Is diabetes increasing, decreasing, or remaining the same in the US population?
- What are some potential causes for the trends in diabetes frequency?

Equipment 

computer with Internet access

LAB 17D OBJECTIVES

- Identify the population trends associated with Type 2 diabetes.

- Identify potential factors that could be causing the increased incidence of Type 2 diabetes.

TYPE 2 DIABETES

This project-based lab activity is intended to lead students to study Type 2 diabetes, which afflicts 95% of all diabetics. While Type 1 diabetes has many of the same effects and symptoms as Type 2 diabetes, the causes and mechanisms of the two diseases are very different.

CLASS DISCUSSION 

Start the lab activity with a class discussion about Type 2 diabetes. What have they heard? What are some factors that might increase the risk of the disease? How is it treated?

ANALYZING DATA SOURCES 

There is a wealth of information available on the Internet. Students should search for information on reputable sites such as the CDC and the NIH. Other good sites would include those of medical and advocacy groups.

NOTES

PRESENTATION
OF NEW KNOWLEDGE 

You may consider different means for students to present what they have learned. They could make a poster, brochure, blog, or even present what they have learned to the class.

CLASS DISCUSSION 

This would be a good opportunity to have a class discussion about the inherent value of all human beings and their physical and medical needs. You might also include a discussion of the importance of taking care of the bodies that God has given us.

Question 1 Answer

Eating a diet low in sugar, consistently exercising, and maintaining a proper body weight make Type 2 diabetes less likely. (*Note*: The causes of Type 2 diabetes can be both genetic and lifestyle based. Some people can make unhealthy choices and never suffer Type 2 diabetes. Others make relatively healthy choices and still deal with this disease. But for those who are genetically predisposed to Type 2 diabetes, healthy lifestyle choices make it less likely that they will develop this condition. Additionally, healthy lifestyle choices often make Type 2 diabetes less severe for those who do develop it.)

Many diabetics must check the amount of glucose in their blood one or more times a day. Some must inject themselves with insulin.

DESIGN YOUR INVESTIGATION

1 Write out the data collection procedures that will allow you to answer the questions that you wrote in Step **4** above. Make sure that your procedures include the importance of obtaining information from reputable sources.

2 Have your teacher approve your procedures.

CONDUCTING YOUR INVESTIGATION

1 Following the procedures that you have written, collect the data needed to answer your questions from Step **4** above.

DEVELOPING A MODEL

1 Analyze your collected data to answer the questions that your lab team wrote about Type 2 diabetes.

2 What changes in the American lifestyle may have contributed to any changes in the number of cases of diabetes?

SCIENTIFIC ARGUMENTATION

1 On the basis of the results of your inquiry, make a claim regarding Type 2 diabetes and support your claim with evidence.

  1. What lifestyle choices make Type 2 diabetes less likely?

    See TE margin for answer.

  2. Read Genesis 1:26–28 and Matthew 22:36–40. How do these passages guide Christians to recognize that the expenses for the care of diabetes are worthwhile?

    The image of God in man and His command to love our neighbor helps Christians recognize that every person has value. For that reason, though diabetes care is expensive, it is worthwhile.

LAB 17E

DO, RE, MI

Gender, Age, and Vocal Range

WHAT IS MY VOCAL RANGE?

Some people have great singing talent while others are completely tone deaf, unable to distinguish one pitch from another. Most people are somewhere in between. But whether or not you can distinguish notes, we can all *produce* different notes. The highest and lowest notes that you can produce are the limits of your *vocal range*.

In this lab activity, you will test how age and gender affect vocal range. You will need to find some volunteers, and the more the better. They should vary in age between seven and seventeen years old, though people outside this range may also participate. Your teacher will give you instructions regarding how to obtain the data, and you and your classmates will work with the same data set.

Procedure

① Use a pitch app to measure the highest and lowest note produced by each volunteer.

NAME:

DATE:

Key Questions
* How does gender affect voice pitch?
* How does age affect voice pitch?

Equipment
pitch app
Excel software

LAB 17E OBJECTIVES

* Determine how gender affects voice pitch.
* Determine how age affects voice pitch.

PITCH APP 

Pitch apps are commonly used by people learning to sing. They identify the note being sung, and many have more elaborate features that you will not need for this lab activity. Many of the simpler apps are available for free and will work quite well.

CALLING ALL VOLUNTEERS 

This lab activity requires a number of volunteers who are older and younger than the students in your class. Possible ways to obtain this data might include working with other science teachers in your school, asking students to recruit their siblings, or finding a more mixed-age group of children, such as a children's church or youth group.

Table 1 is large enough to accommodate data from two children of each gender for each age, for a total of 56 people. But you can complete the activity with far fewer people. At a minimum, you should have at least two girls and two boys younger than twelve, two girls and two boys older than twelve, and a boy and a girl in seventh grade.

It may be helpful to remember that the volunteers do not have to come to any specific place at the same time. Instead, they or their parents or teachers can submit the data to you.

For safety reasons, data points should not be connected to specific children. Only age, gender, and vocal range is required to complete this lab activity.

DO, RE, MI **209**

NOTES

FROM NOTES TO HERTZ

Some pitch apps may identify pitch by the note, such as A_3, G_4, and so on. (Middle C is denoted as C_4.) To translate this notation into hertz, consult an online chart showing the frequency of each musical note. Search the Internet using the keywords "frequency of musical notes."

Question 1 Answer

Answers will vary. Example: Older children will be able to reach lower notes, but younger children will be able to reach higher notes.

② Record the highest and lowest note produced by each volunteer in the appropriate rows and columns of Table 1.

Table 1

Age	Females' Highest Note (Hz)	Females' Lowest Note (Hz)	Males' Highest Note (Hz)	Males' Lowest Note (Hz)
7	525	262	392	196
8	523	220	1047	220
9	659	196	1480	175
10	784	220	1047	165
			880	165
11	880	196		
12	831	208	587	147
13			1245	175
14	587	175	698	196
			1047	165
15			349	117
16			659	98
			523	65
17	880	196		

1. How do you think age will affect how high or low a person can reach? Write your answer as a hypothesis.

   See TE margin for answer.

   ____

2. How do you think gender will affect how high or low a person can reach? Write your answer as a hypothesis.

   Answers will vary. Example: Girls will be able to reach higher notes, but boys will be able to reach lower notes.

3 Use an Excel spreadsheet to plot the data. You should have four sets of data, one for the highest notes for each gender, and one for the lowest notes for each gender.

3. What are the average high and low notes for children below the age of ten?

Answers will vary.

4. What are the average high and low notes for teens above the age of fourteen?

Answers will vary.

5. How did age appear to affect the ability to reach high and low notes?

The ability to reach high and low notes increased with age.

6. Did this agree with your hypothesis? Explain.

Answers will vary.

7. What were the average high and low notes for females?

Answers will vary.

8. What were the average high and low notes for males?

Answers will vary.

9. How did gender appear to affect the ability to reach high and low notes?

Females reached higher notes than males did, while males reached lower notes than females did.

10. Did this agree with your hypothesis? Explain.

Answers will vary.

11. According to your data, around what age does the voice begin to change for girls?

See TE margin for answer.

12. According to your data, around what age does the voice begin to change for boys?

See TE margin for answer.

Question 11 Answer

Answers will vary. Be sure that students' answers are consistent with their data.

Question 12 Answer

Answers will vary. Be sure that students' answers are consistent with their data.

13. What series of physical changes might explain the timing of voice changes?

changes brought on by puberty

14. A five-year-old girl and her twin brother are in a room along with an eighteen-year-old boy and his twin sister. On the basis of your data, who would you expect to be able to sing the highest?

Answers will vary. Ensure that students' answers are consistent

with their data. Most likely it will be the eighteen-year-old girl.

15. On the basis of your data, which of the four people in Question 14 would you expect to be able to sing the lowest?

Answers will vary. Ensure that students' answers are consistent

with their data. Most likely it will be the eighteen-year-old boy.

Going Further

16. Deeper voices result from larger larynxes. How does this explain why men have a protrusion called an *Adam's apple* but women do not?

Answers will vary. The larger larynxes of men protrude while the

smaller larynxes of women usually do not.

FACTORS IN THE ENVIRONMENT

NAME: ___________________

DATE: ___________________

Listed below are three common relationships found in ecosystems. Below the list are some examples of them. In the blank by each example, write the letter of the relationship that is being described. You will use each letter more than once.

Common Relationships in Ecosystems

A. Organisms affect other organisms.

B. Abiotic factors affect organisms.

C. Organisms affect abiotic factors.

___B___ **1.** To keep from becoming too hot, a lizard spends the hottest part of the afternoon in the shadow of a rock.

___C___ **2.** A groundhog digs tunnels in the soil.

___A___ **3.** A squirrel scampers up a tree, sits on a limb, and begins eating a pecan.

___C___ **4.** The roots of a tree have grown into the cracks of a rock. As the roots continue to grow, the rock slowly crumbles.

___A___ **5.** A leech attaches itself to a turtle's flipper and obtains a meal of blood.

___B___ **6.** Birds use stars to navigate during their migrations.

___A___ **7.** A snake swallows a rat.

___B___ **8.** Warm soil and abundant moisture trigger seeds to sprout.

___C___ **9.** When it rains, plants growing on a hillside prevent the water from running down the hill quickly and carrying soil with it.

___B___ **10.** A year of drought causes an oak tree to produce fewer acorns than normal.

REVIEW 18A OBJECTIVES

- Describe the factors that define an ecosystem.

- Distinguish between biotic and abiotic factors in the environment.

NOTES

____

____

____

____

____

REVIEW 18B

BIOMES AND ECOSYSTEMS

NAME:

DATE:

REVIEW 18B OBJECTIVE

- Describe the various types of biomes.

Part 1

From the list below, categorize each biome shown and described.

chaparral	desert	tropical rainforest
coniferous forest	grassland	tundra
deciduous forest	savanna	

1. In this biome it can be either cold or hot, but very little precipitation falls.

   desert

2. Year-round warm temperatures and abundant rainfall are typical here.

   tropical rainforest

3. Mild, wet winters with warm, dry summers are the norm here.

   chaparral

4. Frequent fires and insufficient rain prevent forest growth in this biome.

   grassland

5. This biome is dominated by cone-bearing trees that keep their needles year-round.

   coniferous forest

6. Trees here do not form a closed canopy.

   savanna

7. Much of the ground in this biome remains frozen year-round.

   tundra

8. Trees here shed their leaves each fall.

   deciduous forest

NOTES

Part 2

In the spaces provided, describe the difference between the terms in each pair.

9. ecosystem/habitat

   An ecosystem is an area in which living and nonliving things interact. A habitat is an area within an ecosystem in which an individual organism lives.

10. abiotic factors/biotic factors

    Abiotic factors are the nonliving components of an ecosystem. Biotic factors are the living components.

11. biome/ecosystem

    A biome is a smaller part of the biosphere that shares similar climates and kinds of living things. An ecosystem is a limited area within a biome in which living and nonliving things interact.

12. coniferous forest/deciduous forest

    A coniferous forest is dominated by coniferous (evergreen) trees that keep their needles year-round. A deciduous forest is dominated by trees that lose their leaves each autumn.

13. habitat/niche

    A habitat is the portion of an ecosystem in which an organism lives. An organism's niche is the role it plays within its ecosystem.

BACKYARD ECOSYSTEMS

Inquiring into Ecology Right Outside Your Door

NAME:

DATE:

HOW DO LIVING AND NONLIVING THINGS INTERACT WHERE I LIVE?

How far do you need to travel in order to find a place where living and nonliving things interact? Not very far! You probably have a yard, park, or other open area within walking distance of where you live. Each of these areas has a biotic community (made of various populations of living organisms) and an abiotic environment (made up of nonliving factors). You observe these ecosystems every day. But what specific organisms and factors are found in these ecosystems? And how do they interact with each other?

Procedure

COLLECTING DATA

As you collect data below, record it in Table 1.

1. Choose a limited outdoor area to observe and list the various abiotic factors within it. The area you choose should include factors that humans control. For example, mowing and watering a lawn are factors that are not natural to an area but can be important parts of an ecosystem.

2. List the populations that make up the biotic community of your ecosystem and be as specific as possible. In other words, do not just list "birds"; list the specific kinds of birds that you observe, such as robins. Include both animal and plant populations. Be sure to include both large and small organisms. You may need to use binoculars to observe organisms that won't let you approach closely. You can use a hand lens to observe very small organisms. There may even be organisms in the soil that you can identify.

3. Don't forget that some organisms may use your ecosystem only on a seasonal basis; include these organisms as well. If you have pets that use the area, include them too, even though they may be in the area only part of the time.

Key Questions 

- What kinds of living things are found where I live?
- What kinds of abiotic factors exist where I live?
- What kinds of ecological relationships exist where I live?

Equipment

field notebook
ecosystem to observe
binoculars (optional)
hand lens (optional)

LAB 18C OBJECTIVES

- Identify abiotic and biotic factors in an ecosystem.

- Describe the relationships between biotic and abiotic factors in an ecosystem.

TEACHING TIPS 

This activity is designed to be done outside of classroom time.

No minimum number of observations has been set for this activity. Choose that number on the basis of the interest level and abilities of your students.

This activity can also be done by pairs or small groups.

NOTES

ANALYSIS QUESTIONS 

Consider these four questions to be the minimum response required. At your discretion, have students describe additional relationships that they have observed. They can write these on additional sheets of paper and attach them to this lab activity.

OPTIONAL ASSESSMENTS 

You may also give students options for presenting their findings, such as a display board, PowerPoint presentation, or video.

GOING FURTHER 

If students have access to a camera trap (trail cameras used by hunters are a common example), consider giving them the opportunity to augment their observations with data collected with the camera. Ask students how the camera observations compare with their own field observations.

4. Carefully observe the interactions within your ecosystem. Look to see how organisms interact not only with each other but also with their environment.

5. Choose four interactions from among those that you have observed. One interaction should be between individuals of the same population (such as two robins), another should be between members of two different populations (for instance, between robins and house-cats), and a third should be between an organism and a natural abiotic factor (such as the activity of robins and the amount of daylight present). Your last response should describe the interaction between a population and a manmade abiotic factor (such as how robins interact with a bird bath).

Analysis

1. Write a detailed description of an interaction that you observed between individuals of the same population.

Answers will vary.

2. Describe in detail an interaction that you observed between members of two different populations.

Answers will vary.

3. Describe in detail an interaction that you observed between a population and an abiotic factor.

Answers will vary.

4. Describe in detail an interaction that you observed between a population and a manmade abiotic factor.

Answers will vary.

Going Further

Because of the limited amount of time that you will be able to spend making observations, it is very unlikely that you will see every kind of interaction that takes place in the area you observe. What you need is a device that can make observations when you're not around. One such device is known as a *camera trap.*

5. Do an Internet search using the keywords "camera trap." Look for information about how camera traps are used in the study of ecology. Be sure to take note of the advantages and disadvantages of camera traps. Write a paragraph to summarize what you learn through your research.

Answers will vary. The advantage of a camera trap is that it can collect data over a long period of time without the need of a human observer being present. The disadvantages of a camera trap include the possibility of the trap altering the behavior of the targeted organisms, the need to secure the equipment itself from damage or theft, and the need to secure electronic data, especially if the data is being transmitted offsite.

Table 1

Abiotic Factors	Biotic Community

REVIEW 19A

THE WATER CYCLE

NAME:

DATE:

"Rain, rain, go away!" You probably learned this children's rhyme when you were little. But as you learned in your textbook, rain is an important part of the water cycle. Without rain, lakes would dry up, plants would die, and the entire world would become a desert.

In the illustration below, draw arrows to show the water cycle. Then write the term from the box below in the numbered blank that identifies that feature on the diagram.

clouds	**stream**
evaporation	**water exiting plant leaves**
groundwater	**water entering plant roots**
precipitation	**water seeping through the ground**

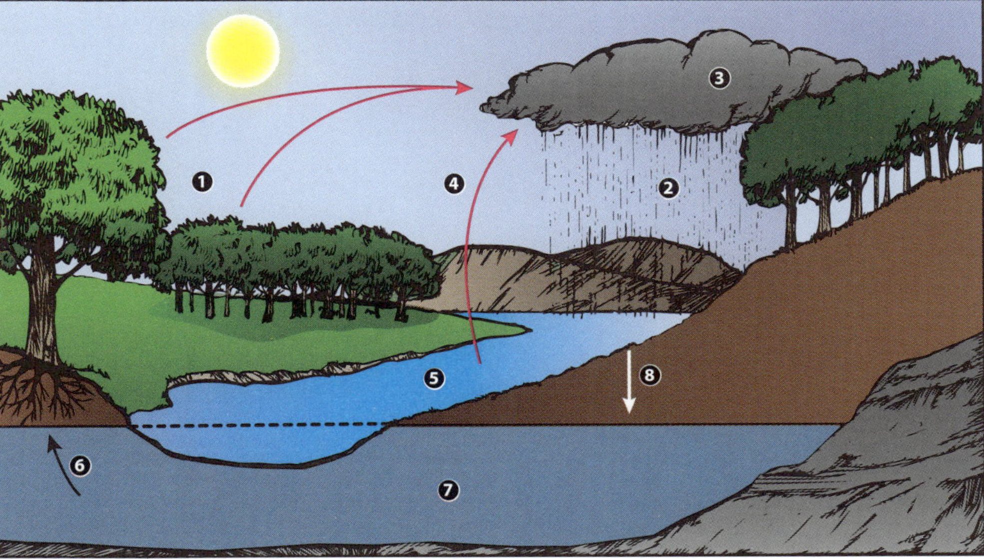

1. water exiting plant leaves
2. precipitation
3. clouds
4. evaporation
5. stream
6. water entering plant roots
7. groundwater
8. water seeping through the ground

REVIEW 19A OBJECTIVE

- Trace the path of the water cycle.

NOTES

MODELING ENERGY FLOW

NAME:

DATE:

Oak-hickory forest ecosystems are common in the Appalachian Mountains of the eastern United States. As you can probably tell from the name, the most common plants are oak and hickory trees. White-tailed deer feast on acorns from oak trees. Other animals, including gray squirrels and black bears, eat both acorns and hickory nuts. Oak borer grubs feed on dead and dying oak trees, and black bears and woodpeckers eat the grubs. Squirrels also eat woodpecker eggs and are eaten by bears, which also occasionally prey on white-tailed deer fawns.

1. Using the information above, create a food web.

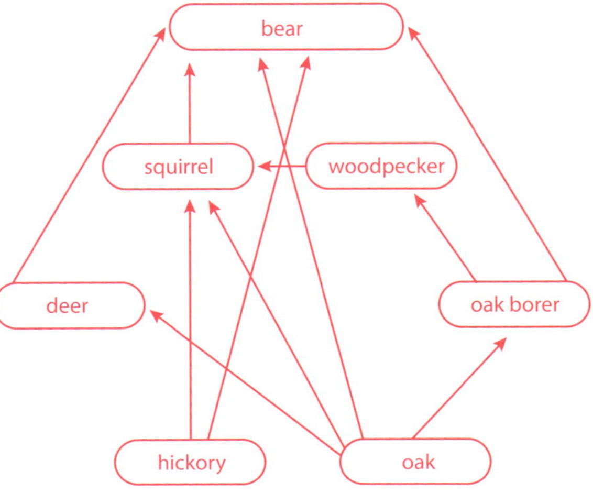

REVIEW 19B OBJECTIVES

- Analyze available energy in an ecosystem.

- Trace the flow of energy through a food web.

NOTES

2. Choose one of the organisms in your food web and create an energy pyramid of a food chain that it is part of.

Answers will vary. An example is shown below.

3. If the organism in the second level of the pyramid contains 100 Calories of energy, how much of that will be available for the animal in the next level?

10 Calories

SUCCESSION ON A VOLCANO

NAME:

DATE:

The Mount St. Helens volcanic eruption on May 18, 1980, carved a huge swath of destruction. Much of the north side of the mountain collapsed in a huge landslide as lava, hot gases, and ash destroyed most of the living organisms around the mountain. Large areas of the forest were flattened, while others were covered by rock made of hardened ash and lava.

Since the 1980 blast, Mount St. Helens has been an excellent place for scientists to study succession. In each question below, an aspect of succession at the site is described. Answer the questions on the basis of what you learned in your textbook about succession.

REVIEW 19C OBJECTIVE

- Classify examples of succession as primary or secondary.

1. In some areas, the blast from the eruption destroyed all life. But within a few months of the eruption, some plants, such as pearly everlasting and fireweed, were found growing in these areas. Is this primary or secondary succession? Explain.

   This is secondary succession. If plants are able to grow within a few months, there must have been soil there.

2. What term describes the pearly everlasting and fireweed in this situation?

   pioneer species

3. Fir trees had grown on the slopes of the mountain before the eruption. Fir seedlings could not regrow in areas that had been covered by volcanic ash and lava. Only after many generations of lichens and mosses and other organisms had grown on the rock and died would firs begin to grow. Is this is an example of primary or secondary succession? Explain.

   This is primary succession. It begins with rock without any soil.

4. What term describes fir trees in this ecosystem?

   climax species

5. The biotic community surrounding Mount St. Helens immediately before the 1980 eruption was not a climax community. It was still recovering from an 1857 eruption. If no more eruptions occur, is Mount St. Helens likely to reach its climax community before 2100? Explain.

   No. Considering that 123 years had not been sufficient for the climax community to recover, it is unlikely that it will do so in 120 years.

6. When the first fir seedlings grow, will the climax community have returned? Explain.

   No. The climax community includes all the species that live in the conifer forest. Seedlings do not provide this environment, so the community will be different.

NOTES

LAB 19D

A TANGLED WEB

My Food Web

AM I A HERBIVORE, CARNIVORE, OR OMNIVORE?

On a summer's day, there's nothing like a grilled hamburger with tomato, lettuce, and pickles on it. Sounds good, doesn't it? Foods from many different organisms come together to make a delicious sandwich. Tomato plants, lettuce plants, cucumber plants, wheat seeds, and a cow are all part of that meal. That burger is a small food web in itself!

In this activity, you will list and analyze the foods in several meals to discover the organisms that you eat. Hopefully this won't make you too hungry!

Procedure

1. In the *Dishes* column of Table 1, write the name of individual dishes from three meals that you have eaten in the past few days.

2. Write the main components of each dish in the *Components* column.

3. Write the organisms that produced each component in the *Organisms* column.

4. In the *Food Chain* column, list the organisms involved in a food chain that ends with each organism that you wrote in the *Organisms* column. Many animals eat a variety of things, so take your best guess. If the organism is a producer, write "none."

1. From your list of foods, which might a herbivore eat?

   Answers will vary. Examples: lettuce, carrots, celery, tomatoes, wheat, peppers, onions, zucchini, apples, grapes, olives, eggplant, garlic, pecans

2. From your list of organisms, which might a carnivore eat?

   Answers will vary. Examples: cow, pig, chicken

3. Using your answers to Questions 1 and 2, how would you be classified on the basis of your diet?

   See TE margin for answer.

NAME:
DATE:

Key Questions 

- What organisms produce my food?

- What organisms are part of my food web?

Equipment E

none

LAB 19D OBJECTIVES

- List the organisms that produce items of food.

- Classify the organisms within a food chain and food web.

UNCONSUMED ORGANISMS 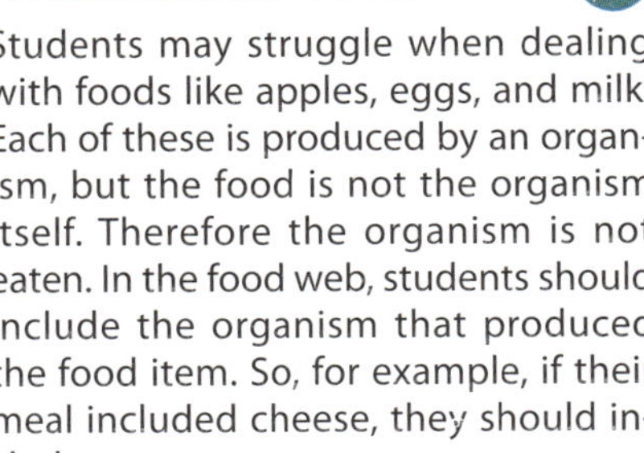

Students may struggle when dealing with foods like apples, eggs, and milk. Each of these is produced by an organism, but the food is not the organism itself. Therefore the organism is not eaten. In the food web, students should include the organism that produced the food item. So, for example, if their meal included cheese, they should include a cow.

Question 3 Answer

an omnivore (A vegetarian student could correctly claim to be a herbivore.)

NOTES

5 Using the information that you wrote in Table 1, create a food web of the organisms that are involved in your diet.

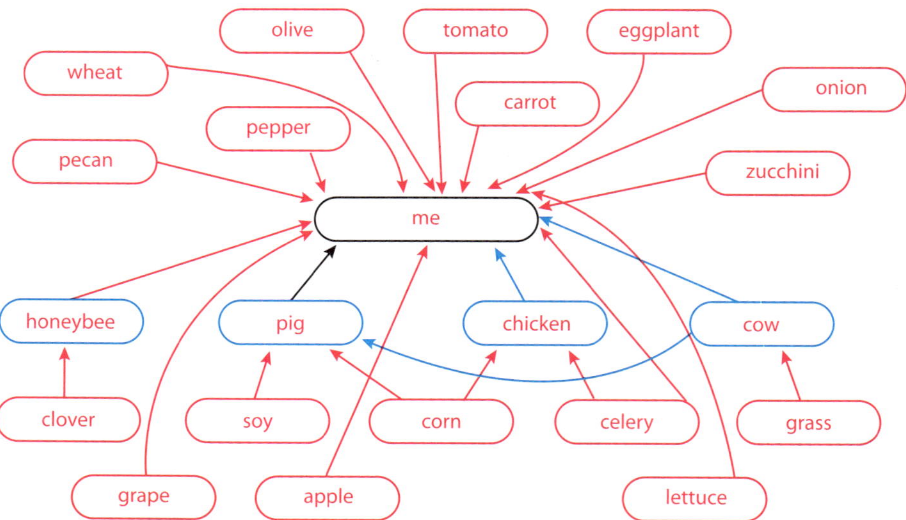

human or animal eats producer
human or animal eats herbivore
human or animal eats carnivore or omnivore

olive
tomato
eggplant
wheat
carrot
onion
pepper
pecan
zucchini
me
honeybee
pig
chicken
cow
clover
soy
corn
celery
grass
grape
apple
lettuce

NAME: _______________

Dishes	Components	Organisms	Food Chain
garden salad	lettuce	lettuce plant	none
	carrots	carrot plant	none
	celery	celery plant	none
	tomatoes	tomato plant	none
pizza	dough	wheat	none
	cheese	cow	grass, cow
	green peppers	pepper plant	none
	onions	onion plant	none
	pepperoni	pig	soy, corn, cow, pig
zucchini casserole	zucchini	zucchini plant	none
	bread crumbs	wheat	none
	cream of chicken soup	chicken	corn, chicken
	sour cream, butter	cow	grass, cow
baked chicken	chicken	chicken	corn, chicken
apple salad	apples	apple tree	none
	raisins	grapevine	none
	celery	celery plant	none
	yogurt	cow	grass, cow
Greek salad	lettuce	lettuce plant	none
	tomatoes	tomato plant	none
	banana peppers	pepper plant	none
	Kalamata olives	olive tree	none
	feta cheese	cow	grass, cow
moussaka	ground beef, milk	cow	grass, cow
	eggplant	eggplant plant	none
	tomato sauce	tomato plant	none
	garlic	garlic plant	none
	onion	onion plant	none
baklava	dough	wheat	none
	pecans	pecan tree	none
	honey	honeybee	clover plants, honeybee

SAMPLE DATA ✓

The data in Table 1 is sample data. Your students' data may be different.

Going Further

4. Some people argue that the widespread adoption of a vegetarian life-style is necessary to feed the world's growing human population. Can that argument be supported on the basis of what you have learned about food webs and energy pyramids? Explain.

If humans eat only plant-based foods, that is, act as the primary consumers on an energy pyramid, then the number of humans that can be sustained by a given amount of primary production is greater than when humans act as secondary or tertiary consumers. But many of the animals that people eat live on grass, which humans cannot digest. So eating meat allows humans to gain some energy indirectly from plants that they would otherwise be unable to gain any energy from. Furthermore, meat contains some nutrients that are difficult to obtain in a plant-based diet.

5. Does the Bible give us any information about whether a vegetarian diet is preferable to one that includes meat?

Answers will vary. In the original Creation Mandate, God gave every green plant as food for man. After the Flood, God gave man "every moving thing" (Gen. 9:3) as food in addition to plants. Subsequently, God placed certain restrictions on what could be eaten by His covenant people, Israel, but He never forbade the eating of meat.

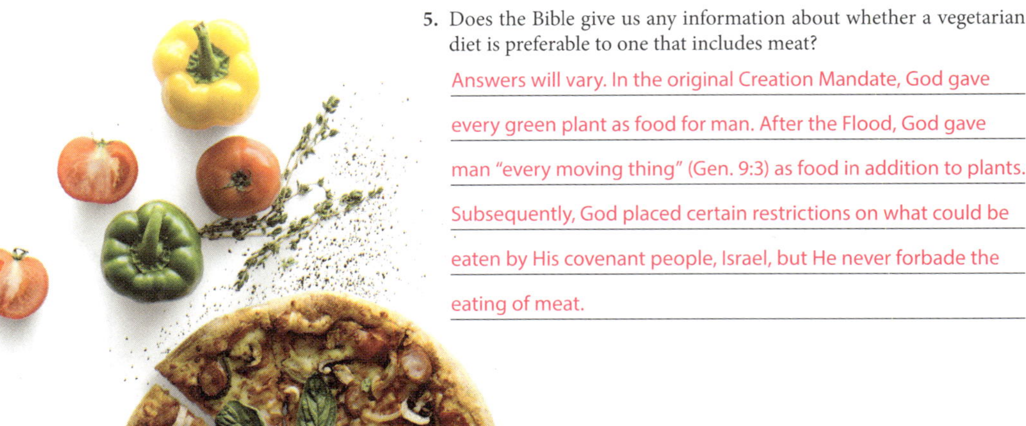

THE OLD, ABANDONED FIELD

Observing Succession

NAME:

DATE:

WHAT DOES SUCCESSION LOOK LIKE?

Have you ever seen a field like the one shown below? No crops have been planted in it for several years, and now brambles and brush are starting to grow. Soon small trees will spring up. Eventually, this field will look exactly the same as the surrounding forest.

As you learned in your textbook, this process is called *ecological succession*. It takes a long time, often many decades. In this activity, you will visit a site where succession is occurring. Observe closely and you will learn more about this amazing process.

Key Questions

- Is a given area experiencing primary or secondary succession?
- What plants are part of the different stages of succession?

Equipment

field guides or keys

Procedure

1 Observe the plants in the center of the field and those at the edge.

1. Are the plants in the center of the field different from those at the edge?

See TE margin for answer.

2. Is there a nearby area with a climax community? Explain.

See TE margin for answer.

3. If you answered "yes" to Question 2, what are the primary plant species of the climax community?

Answers will vary.

LAB 19E OBJECTIVES

- Determine whether an area is experiencing primary or secondary succession.

- Identify the plants in different stages of succession.

SUITABLE SITES 

This lab activity requires you to take your class on a field trip. If you live in or near a rural area, a field that has not been disturbed for at least several years is ideal. If you are in an urban area, an abandoned lot would work adequately.

If you have several options, choose one where you can obtain some background information about when the area was last used.

TAILOR THE LAB 

Succession as it occurs in a natural setting is often a very complex process. Students will likely find the task of analyzing it challenging. Modify this valuable lab activity to best fit your particular class.

CLIMAX COMMUNITY? 

Sometimes an intermediate stage of succession can look like a climax community. Contact a local forester to learn the climax communities of your area.

Question 1 Answer

Answers will vary. If the area is surrounded by an area of climax vegetation, it's likely that the edges will be in a more advanced stage of succession than the center.

Question 2 Answer

Answers will vary. Old fields are often near forested or grassland regions. Abandoned urban lots will generally not be near an area with a climax community.

NOTES

❷ Identify and write the names of three pioneer species in the *Pioneer Species* column of Table 1.

Table 1

Pioneer species	Intermediate species	Climax species
Timothy grass	black locust	red oak
goldenrod	sassafras	shagbark hickory
purple aster	autumn olive	sugar maple

❸ Identify and write the names of three plant species that appear to be neither pioneer species nor climax species in the *Intermediate Species* column.

❹ If any plants from the climax community are growing in the area, use your field guide to identify three of them and write their names in the *Climax Species* column.

4. On the basis of the number of pioneer, intermediate, and climax species, is the succession of this location closer to the beginning or the end?

Answers will vary.

5. Make an estimate of the number of years that have passed since succession began.

Answers will vary.

6. Search for records of the area or for someone who has lived in the area for a long time. How many years has succession been occurring?

Answers will vary.

7. Was your estimate in Question 5 close?

Answers will vary.

8. On the basis of your research, what did the area look like before succession started?

See TE margin for answer.

9. Is this an example of primary or secondary succession? Explain.

Answers will vary. Most areas will be examples of secondary succession.

Going Further

Succession often occurs when an area is abandoned by people. Choose an area that has experienced a significant decrease in population in the last 200 years. Do some research on why the people left. Compare your research with that of your classmates.

NAME: _______________________

10. Using the research that you and your classmates have done, what do you think are some reasons people abandon areas?

Answers will vary. People often abandon an area for economic reasons (e.g., Kolmanskop, Namibia) as well as ecological (e.g., Plymouth, Montserrat) or political (the Varosha section of Famagusta) reasons.

COLLABORATIVE LEARNING: POOLING INFORMATION

The *Going Further* section will allow students to pool independently gathered research in order to make some generalizations about why people abandon an area.

THE OLD, ABANDONED FIELD **233**

REVIEW 20A

POLLUTION

NAME:

DATE:

When some people think about pollution, they think about black, oily, tarry substances. But pollution can also look like what you see in the image at right. As you learned on page 421 of your textbook, bright lights can be a form of energy pollution.

In Questions 1–5, classify each place using the following:

A. energy pollution

B. biodegradable substance pollution

C. nonbiodegradable substance pollution

D. no pollution

REVIEW 20A OBJECTIVE

- List the various forms of pollution.

1. _____ B

2. _____ C

3. _____ A

4. _____ D

5. _____ C

NOTES

DIFFERENT ANSWERS POSSIBLE ✓

The distinction between different ways of dealing with wastes is not always clear. Some students might argue that since the lumber companies are taking a waste and turning it into something useful, they are recycling it. That answer is acceptable, though it can also be argued that since they are turning what once was a waste into a profitable product, they are reducing the amount of waste.

In Questions 6–9, classify each example of dealing with wastes using the following:

A. reduction

B. recycling

C. treatment

D. safe disposal

_____ C   **6.** A solution of sodium hydroxide is carefully mixed with a hydrochloric acid solution. The result, salt water, is then poured down the drain.

_____ A   **7.** Many lumber mills once burned waste wood chips and saw dust, producing large amounts of smoke. Now lumber mills use their wood wastes to make products such as particle board and mulch.

_____ D   **8.** To prevent trash from piling up, a city collects its citizens' trash once a week. It then takes the garbage to the local landfill, where the trash is buried in carefully designed pits.

_____ B   **9.** A landscaping company collects large amounts of leaves. They used to dump them on a piece of land by a stream, but the leaves began causing problems in the stream. Now the company sells the leaves to another company that makes compost from them.

REVIEW 20B

RENEWABLE AND NONRENEWABLE RESOURCES

NAME:

DATE:

REVIEW 20B OBJECTIVE

- Contrast renewable and nonrenewable resources.

With skill gathered from years of experience, a woodcutter brings down his ax on a log, which splits into two pieces of firewood. In addition to preparing for winter, this woodcutter is utilizing both renewable and nonrenewable resources. The firewood and the wooden ax handle are made from renewable resources while the steel ax head is made from a nonrenewable resource.

In Questions 1–12, write an *R* if the resource is renewable or an *N* if it is nonrenewable.

__N__ **1.** aluminum   __N__ **7.** natural gas

__N__ **2.** iron ore   __N__ **8.** petroleum (oil)

__R__ **3.** cotton   __N__ **9.** uranium

__R__ **4.** deer   __R__ **10.** trout

__N__ **5.** gold   __R__ **11.** water

__R__ **6.** trees   __R__ **12.** wool

RENEWABLE AND NONRENEWABLE RESOURCES   **237**

NOTES

In Questions 13–15, each statement presents a problem in managing natural resources along with a solution. (1) Indicate whether the solution presented fits more with a conservation or preservation philosophy of resource management. (2) Then suggest how people who hold to the other philosophy would have dealt with the problem.

13. After the previous two deer hunting seasons, the population of deer had decreased dramatically. State officials reduce the number of deer that each hunter is allowed to harvest from two to one.

(1) conservation

(2) Answers will vary. Example: Preservationists would have eliminated the hunting season altogether.

14. A newly discovered species of tree exists in one small area. The area's owner sets it aside as a trust so that the trees can never be cut down.

(1) preservation

(2) Answers will vary. Example: conservationists would be willing to carefully use the trees at some point.

15. The number of teak trees being cut down will soon deplete the supply of naturally growing teak trees. Teak companies begin planting a large number of young teak trees each year to replace the trees that they cut down.

(1) conservation

(2) Answers will vary. Example: Preservationists would have stopped cutting teak altogether.

MANAGING THE ENVIRONMENT

NAME:

DATE:

In the Creation Mandate, God gave humans the privilege and responsibility of caring for the earth and its natural resources. Many times, however, we have failed to obey God's command. The curse that resulted from the Fall makes obeying the Creation Mandate even harder. But the careful use of sound management principles can enable us to make the best use of God's creation.

On page 433 of your textbook, you learned about the management principles of maximizing usefulness, minimizing hidden costs, and maximizing the long-term health of resources. Read the case study that follows, then answer Questions 1–6.

AGRICULTURAL PRACTICES IN MADAGASCAR

Famers in Madagascar often use a method called *slash-and-burn*. They move into an area of tropical forest, cut the trees, and then burn them. The ash provides extra nutrients to the soil where the farmers then grow their crops. After a year or two, the nutrients in the soil are used up, so the farmers move on to another patch of forest. After a few years, they may return to the original patch and start the process over again.

Over the long term, slash-and-burn farming creates many problems. Often an area of land is not given enough time to build up nutrients before the farmers plant more crops, so the forest never really regrows. Eventually, the lack of trees allows water to wash away any remaining nutrients. The lack of nutrients also results in poor crop yields. Over decades and centuries, much of Madagascar's original lush forests have been turned into dry grasslands that do not support most of the animal life or make good cropland.

REVIEW 20C OBJECTIVE

- Apply the principles for wisely managing God's world to specific examples.

AGRICULTURE IN A MODERN URBAN NATION

For an article on agricultural success in the Netherlands, a densely populated, urbanized, and industrial nation, conduct an Internet search for "This Tiny Country Feeds the World," originally published in the September 2017 issue of *National Geographic*®.

NOTES

WHAT IS BETTER? !?!

It's important to note that better farming methods in the context of Madagascar may not be the farming methods typically employed in the United States. Each country and region has its own culture, needs, and climate that create agricultural circumstances that may require unique solutions in order to allow the farmers to practice sustainable agriculture. But learning what some of those solutions might be for Madagascar would be an in-depth study and far beyond the scope of this course.

1. How does slash-and-burn farming fail to maximize the usefulness of natural resources?

   Slash-and-burn farming removes nutrients from the soil without putting any back. This causes the land to become less useful for both humans and wildlife.

2. The farmers who practice slash-and-burn farming are attempting to provide food for themselves and their families. But what are some of the hidden costs of their methods?

   Slash-and-burn farming destroys much of the forest in Madagascar. This reduces forest resources, including trees and animals. The dry grasslands do not make good cropland, so slash-and-burn reduces the amount of farmland available.

3. How does slash-and-burn farming fail to maximize the long-term health of resources?

   Resources of the land include the animals, the trees, and the nutrients in the soil. Slash-and-burn agriculture depletes the soil nutrients. It destroys the forests, depriving the animals of their habitat. It also reduces the ability of the soil to support more crops. Eventually, this process will destroy all three resources.

4. Farmers in many countries use the same plot of ground year after year. To avoid exhausting the soil nutrients, farmers practice crop rotation and plant cover crops, and they add fertilizer to their soil. How would these techniques improve the usefulness of Madagascar's natural resources for both people and animals?

   Agricultural practices from developed countries would allow Madagascan farmers to get more crops from a much smaller area of land. So farmers could grow their crops in one place rather than move every few years. Finally, more areas of the country would remain forest, benefiting the wildlife of the island.

5. How would better farming methods help avoid the hidden costs of slash-and-burn farming?

   Modern farming methods require smaller, permanent tracts of land, allowing more land to remain forested, preserving the nutrients in the soil.

6. How would better farming methods help maximize the long-term health of the resources?

   Modern farming methods would allow farming to continue for the long term while also allowing other resources, such as trees and animals, to thrive.

LAB 20D

CLEAN AND BRIGHT

Recycling Paper

IS RECYCLED PAPER USEFUL?

Take a look at several paper items near you, such as a textbook, a note-book, or note cards. Do any of them indicate that they are made from recycled paper? Most cities and counties have centers for receiving recy-clable materials. The materials are transported from these centers to facilities that turn the old materials into new products.

In this activity you will recycle some paper. While paper is normally recycled on a much larger scale than what you will use, the processes are very similar.

Procedure

TURNING PAPER INTO PULP

1. Obtain the screen frame provided by your teacher. It should be smaller than the dishpan.

2. Cut or tear an amount of paper equal to a two-page spread of a news-paper into approximately 2 cm squares.

3. Place the paper into a blender with 1 L of water, and blend it until it becomes a cloudy liquid (15–30 s). This turns the paper back into pulp.

1. How do you think the pulp that you made differs from the pulp used to make brand-new paper? How is it similar?

   Brand-new pulp is made directly from wood. Recycled pulp is made from paper and is thus indirectly made from wood. Both are wood products.

4. Pour the pulp into the dishpan. Continue making pulp until the dishpan is filled to within 8 cm of the top.

5. Stir the pulp in the dishpan with your hands. You may need to add water if the pulp seems thick.

NAME: _______________

DATE: _______________

Key Questions 

- How is paper recycled?
- Is it possible to recycle paper in a classroom?

Equipment

wooden frame with mesh screen

paper to be recycled (newspaper)

blender

water

large dishpan

piece of cloth (or blotting paper)

large sponge

heavy object (such as a large book)

LAB 20D OBJECTIVES

- Make recycled paper.

- Model on a small scale how paper is recycled.

MAKING THE WOODEN FRAME 

To make the frame, obtain a 10 ft 1 × 4 board. Cut the lumber into the follow-ing lengths:

$$18 \text{ in. (2 pcs.)}$$
$$15 \text{ in. (2 pcs.)}$$
$$11 \text{ in. (2 pcs.)}$$
$$8 \text{ in. (2 pcs.)}$$

Obtain a 16 in. × 13 in. piece of metal window screen.

Arrange the 18 in. and 8 in. pieces flat on the ground to make a rectangular frame that is 18 in. × 15 in. Lay the screen on top so that the edges of the screen are approximately 1 in. from each outside edge of the frame. Staple the screen to the frame. Lay the 15 in. and 11 in. pieces on top and use 1¼ in. screws to join the upper and lower frames.

DISHPAN 

Because the frame will be 18 in. × 15 in., the dishpan should be larger than this.

CLEAN AND BRIGHT **241**

NOTES

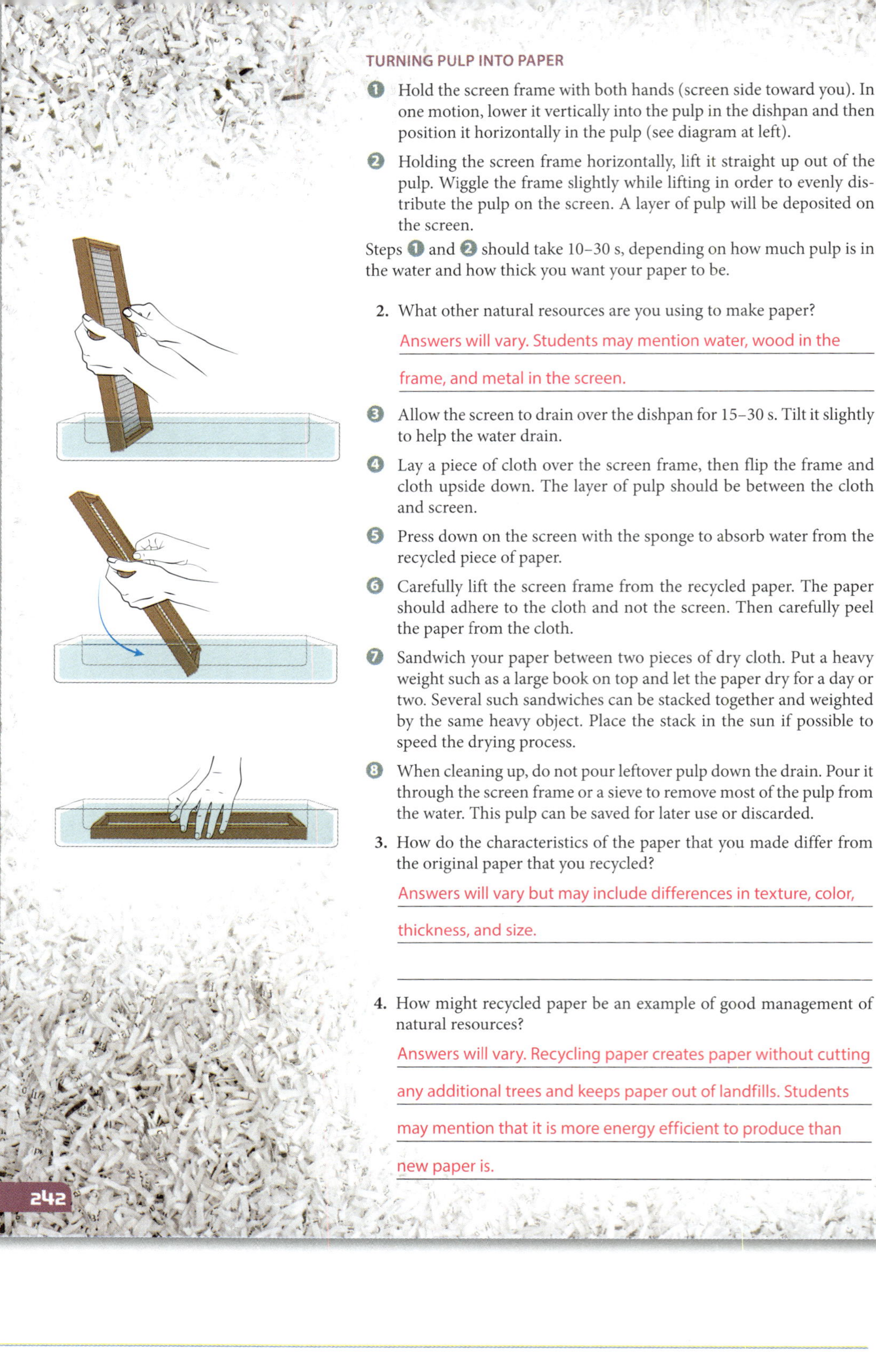

TURNING PULP INTO PAPER

1. Hold the screen frame with both hands (screen side toward you). In one motion, lower it vertically into the pulp in the dishpan and then position it horizontally in the pulp (see diagram at left).

2. Holding the screen frame horizontally, lift it straight up out of the pulp. Wiggle the frame slightly while lifting in order to evenly distribute the pulp on the screen. A layer of pulp will be deposited on the screen.

Steps 1 and 2 should take 10–30 s, depending on how much pulp is in the water and how thick you want your paper to be.

2. What other natural resources are you using to make paper?

Answers will vary. Students may mention water, wood in the

frame, and metal in the screen.

3. Allow the screen to drain over the dishpan for 15–30 s. Tilt it slightly to help the water drain.

4. Lay a piece of cloth over the screen frame, then flip the frame and cloth upside down. The layer of pulp should be between the cloth and screen.

5. Press down on the screen with the sponge to absorb water from the recycled piece of paper.

6. Carefully lift the screen frame from the recycled paper. The paper should adhere to the cloth and not the screen. Then carefully peel the paper from the cloth.

7. Sandwich your paper between two pieces of dry cloth. Put a heavy weight such as a large book on top and let the paper dry for a day or two. Several such sandwiches can be stacked together and weighted by the same heavy object. Place the stack in the sun if possible to speed the drying process.

8. When cleaning up, do not pour leftover pulp down the drain. Pour it through the screen frame or a sieve to remove most of the pulp from the water. This pulp can be saved for later use or discarded.

3. How do the characteristics of the paper that you made differ from the original paper that you recycled?

Answers will vary but may include differences in texture, color,

thickness, and size.

4. How might recycled paper be an example of good management of natural resources?

Answers will vary. Recycling paper creates paper without cutting

any additional trees and keeps paper out of landfills. Students

may mention that it is more energy efficient to produce than

new paper is.

5. How might the large-scale production of recycled paper cause pollution?

Answers will vary. One potential problem is that pulp not filtered from the water could pollute water supplies.

6. What recycled paper products have you seen in use?

See TE margin for answer.

Going Further

7. Most writing paper is white. Why might this be an issue with recycling paper?

Answers will vary. Students should mention that much of the paper being recycled has been treated with inks and dyes, which would discolor the recycled paper.

8. Perform an Internet search for the word "de-inking." What are some common de-inking processes?

The de-inking process removes ink from paper pulp during the recycling process. Flotation de-inking uses chemicals to remove the ink from the pulp and then carry it to the surface of the water to be washed away. Bleaching de-inking uses chemicals to break down the inks, whitening the paper. Washing uses large amounts of water to remove inks.

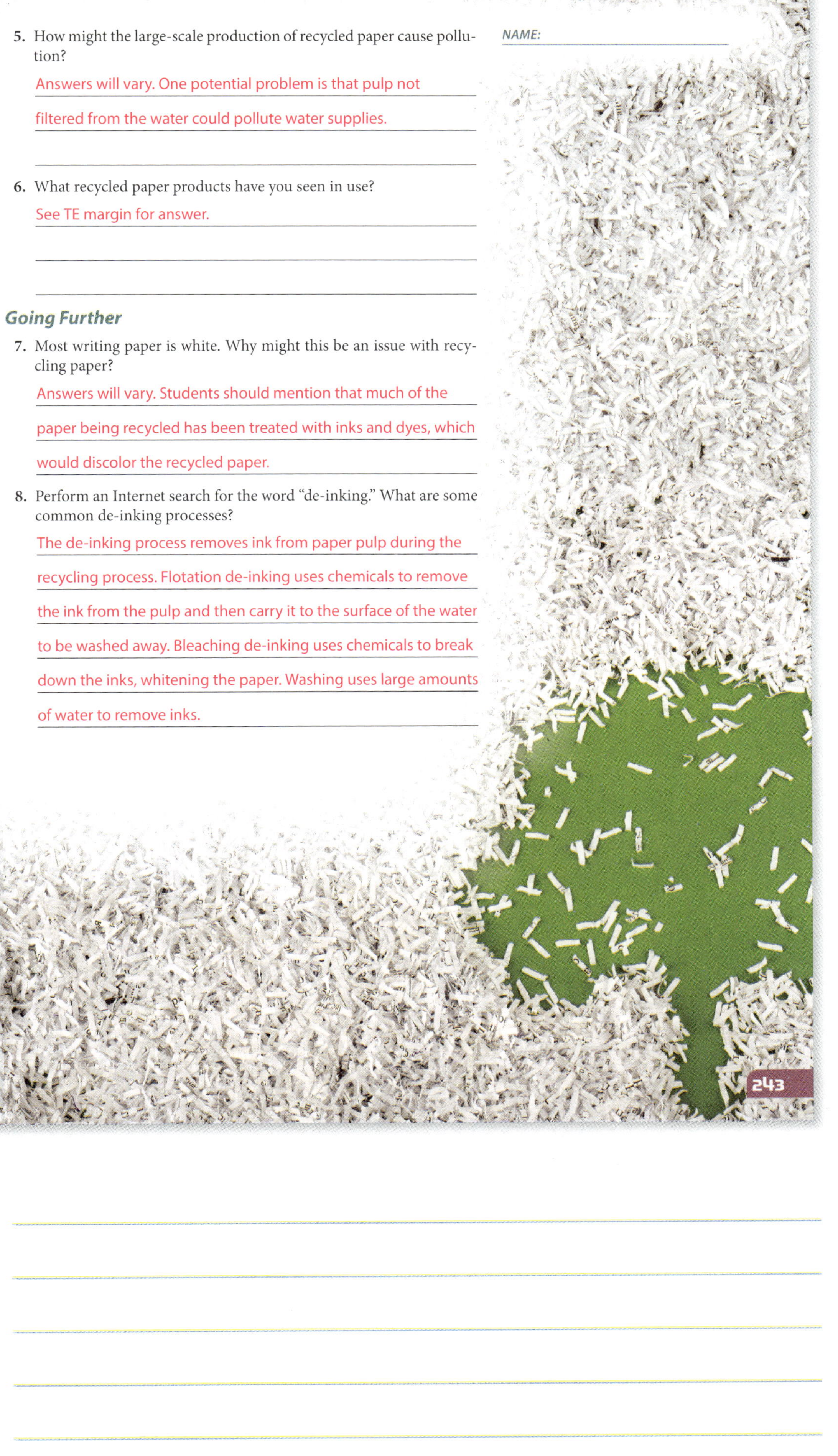

Question 6 Answer

Answers will vary but will generally include paper items, such as bags, cardboard boxes, and envelopes. Students may also mention construction materials, such as insulation and certain types of wood substitutes.

LAB 20E

POPULATION EXPLOSION

Modeling Population Growth

HOW MANY HUMANS LIVE ON THE EARTH?

People who are interested in the world's population aren't only interested in the population in years past. They are even more interested in predicting the number of people in future years. They use models to predict the number of people who will live on the earth in 2050, 2100, and even further into the future. In the graph below, the dashed line is an *extrapolation*, a prediction that goes beyond the data. But a model can be wrong. Models usually work best when making short term predictions.

NAME:

DATE:

Key Questions

- How can a graph be used to estimate future trends?
- Are long-term extrapolations less accurate than short-term ones?

Equipment

none

LAB 20E OBJECTIVES

- Use a graph to extrapolate into the future.
- Test whether extrapolation becomes less reliable as time increases.

POPULATION EXPLOSION   245

NOTES

246 LAB 20E

Procedure

1 Use the graph on page 245 to complete the *Graph Predictions* column of Table 1.

2 In the *Data or Extrapolation?* column, indicate whether the number you wrote was a data point or an extrapolation.

3 For each data point you marked as an extrapolation, calculate its percent error using the data in the *UN Population Figures* column as the accepted value. Percent error is calculated using the formula below.

$$\text{percent error} \;=\; \frac{|\text{model prediction} - \text{accepted population}|}{\text{accepted population}} \times 100\%$$

4 Record your data in the *Percent Error* column.

1. How accurate were the extrapolations? (A percent error less than 5% is considered reasonably accurate.)

   Answers will vary.

2. What factors will influence the rate at which the world population increases?

   Any changes in the birth or death rates will affect the rate at which the population changes.

Table 1

Year	Graph Prediction (billions)	Data or Extrapolation?	UN Population Figures (billion)	Percent Error
1975	4.1	data	4.1	
1987	5.1	data	5.1	
1998	6.0	data	6.0	
2004	6.5	data	6.5	
2010	6.8	extrapolation	7.0	2.9%
2015	7.2	extrapolation	7.4	2.7%
2017	7.3	extrapolation	7.6	3.9%

POPULATION FIGURES 

Some students may wonder where the numbers in Table 1 come from. It is impossible to count every person in the world or even in the United States. Therefore all large population figures are estimates. The challenge of obtaining an accurate count is compounded by the fact that the population is constantly changing as people are born and die every minute.

The answers given in the *Graph Prediction* column may be difficult for students to estimate exactly from the graph. Most students should be able to obtain an answer within 0.1 billion of the answer given.

A slightly different answer will also change the percent error.

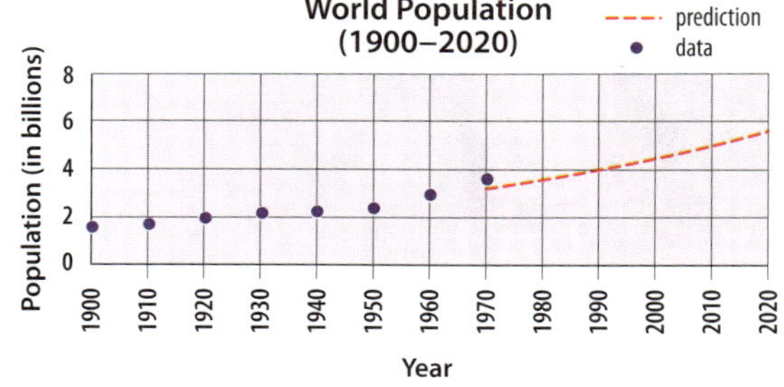

Derived from https://en.wikipedia.org/wiki/World_population_estimates (Hyde column)

Going Further

If the data used is accurate, short-term extrapolations will often be fairly accurate. But will the accuracy of extrapolations decrease as predictions are made further into the future?

3. Extrapolations from data from 1900 to 1970 predicted that the world population in 2010 would be 5.4 billion people. Using the UN population figure in Table 1 for 2010, determine the percent error for this prediction.

   22.9% error

4. Was the prediction accurate?

No. The percent error is greater than 5%.

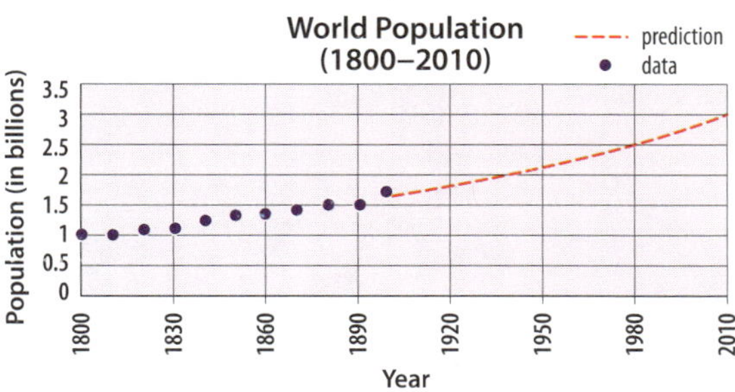

Derived from https://en.wikipedia.org/wiki/World_population_estimates (Hyde column)

5. Extrapolations from data from 1800 to 1900 predicted that the world population in 2010 would be 3.1 billion people. Using the UN population figure in Table 1 for 2010, determine the percent error for this prediction.

55.7% error

6. Was the prediction accurate?

No. The percent error is far greater than 5%.

7. What does this indicate about the accuracy of extrapolations the further they are made into the future?

Long-term extrapolations are often much less accurate.

8. What are some possible reasons for this?

Answers will vary. Much of the inaccuracy is caused by the fact

that extrapolations assume that all factors will remain constant.

While this is a relatively safe assumption in the short term, it is

often not so over the long term. Many factors may change in

ways that extrapolations cannot predict.

9. Name two factors that affect population growth and are hard to predict.

See TE margin for answer.

Question 9 Answer

Answers will vary. Examples include war, famine, abundance of food, natural disasters, and the ability to transport needed food and water from areas of abundance to areas of need.

LABORATORY AND FIRST-AID RULES
LABORATORY RULES

1. Never perform an unauthorized experiment or change any assigned experiment without your teacher's permission.

2. Avoid playful, distracting, or boisterous behavior.

3. Work at your own lab station.

4. Always wear safety goggles when working with chemicals and other materials or objects that are potentially hazardous to the eyes.

5. Wear protective clothing and gloves when working with corrosive or staining chemicals.

6. While working in the laboratory, tie back long hair and avoid wearing loose clothing such as scarves.

7. Never taste any chemical or eat or drink out of glassware found in the laboratory.

8. Always use recommended instruments for cutting and handle them carefully. Always cut away from yourself.

9. When handling live organisms, follow instructions and do not cause them undue harm or discomfort.

10. Thoroughly wash your hands with soap after handling any live organisms or cultures containing organisms.

11. To smell a substance, gently fan its vapor toward you.

12. Never leave a flame or heater unattended. Keep combustible materials away from sources of heat.

13. When diluting acid solutions, always add the acid to water slowly. *Never add water to an acid!*

14. When heating a test tube, point the open end away from you and others. *Never heat a closed or stoppered container!*

15. Dispose of waste as instructed by your teacher.

16. Do not return unused chemicals to a container. Dispose of them properly.

17. Notify the teacher of any injuries, spills, or breakages.

18. Know the locations of the fire extinguisher, safety shower, eyewash station, fire blanket, first-aid kit, and Safety Data Sheets.

Appendix A

1. **Burns** Flush the area with cold water for several minutes.

2. **Chemical spills** Notify your teacher of all chemical spills.

   a. On a laboratory desk

      1. If the material is not particularly volatile, toxic, or flammable, your teacher may have you clean the spill. For liquids, use an absorbent material that will soak up the chemical. For solids, use the designated dustpan and brush. Dispose of chemicals and cleaning materials properly. Then clean the area with soap and water.

      2. If the material is volatile, toxic, or flammable, and not a large spill, ask your teacher for help. If it is a large spill, you may need to evacuate the laboratory.

      3. If highly reactive materials such as hydrochloric acid are spilled, your teacher will clean them up.

   b. On a person

      1. If the spill covers a large area, remove all contaminated clothing while under the safety shower. Flood the affected body area for fifteen minutes. Obtain medical help immediately.

      2. If the spill covers a small area, immediately flush the affected area with cold water for several minutes. Then wash the area with a mild detergent solution.

      3. If the spill is an acid, rinse the area with sodium bicarbonate solution; if it is a base, use boric acid solution.

      4. If the chemical splashes in the eyes, immediately flush the eyes in the nearest eyewash fountain for several minutes. Get medical attention.

3. **Fire**

    a. For any fire other than a contained fire, do *not* attempt to put it out on your own. *Understand that some fires can't be put out with water.*

    b. Smother a small fire in a container by covering it.

    c. If a person's clothes are on fire, remember to stop, drop, and roll—roll the person on the floor and use a fire blanket to extinguish the flames. The safety shower may also be used. ***Do not use a fire extinguisher.***

4. **Swallowed chemicals**

    Determine the specific substance ingested. Contact the Poison Control Center in your area immediately.

5. **Cuts**

    Clean cuts with a disinfectant such as hydrogen peroxide or alcohol. Apply triple antibiotic ointment and cover with a bandage if the wound is superficial. If it is a deep cut, seek immediate medical attention.

6. **Bites and stings**

    For minor bites (that do not break the skin) and stings, wash the area with soap and water and cover with antibiotic ointment and a clean bandage.

    If the bite is deep or if the animal is suspected of carrying rabies, see a doctor immediately. The affected area should also be monitored for infection.

7. **Rashes from plants**

    Rashes caused by exposure to poison ivy, oak, or sumac will not appear immediately after contact; it usually takes about twenty-four hours for symptoms to show. If you suspect that you have come into contact with these plants, wash the area with soap and water within ten minutes of contact. Be sure to clean any clothing that has come into contact with plants since it may also contain the toxins.

    If a rash is visible, treat with calamine lotion, hydrocortisone cream, or oral antihistamines (or a combination of the three). If conditions worsen, the affected person should see a doctor.

Appendix B

LABORATORY EQUIPMENT

beaker

concavity slides

graduated cylinder

petri dish

safety goggles

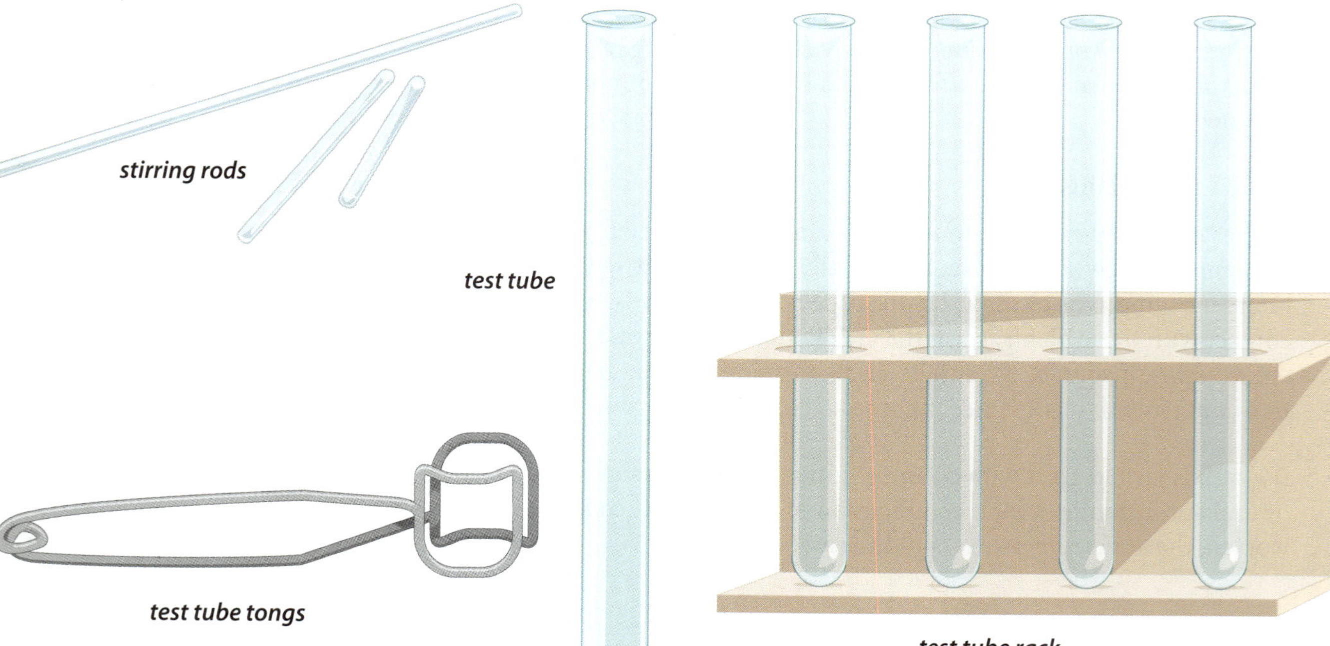

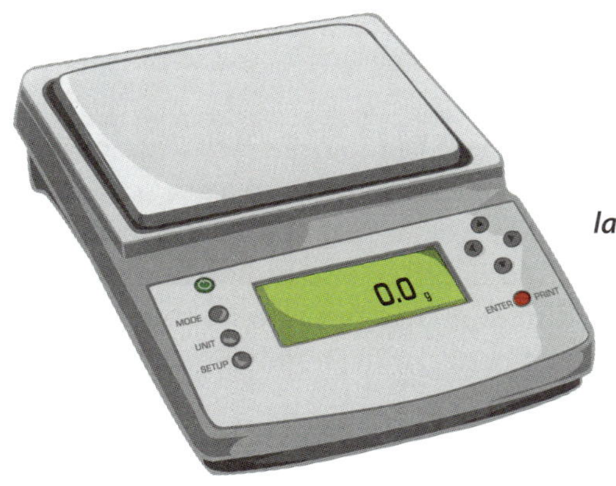

Appendix B

laboratory balance (electronic)

laboratory balance (triple beam)

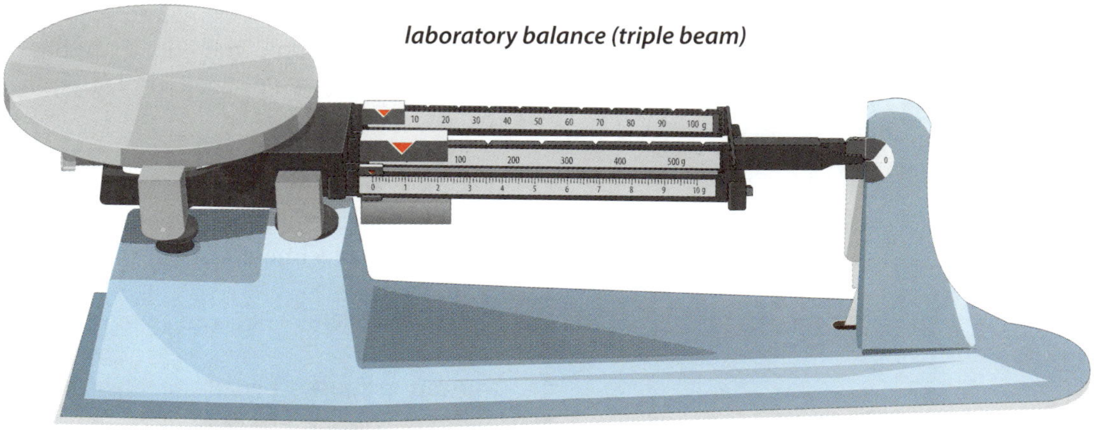

disposable pipette

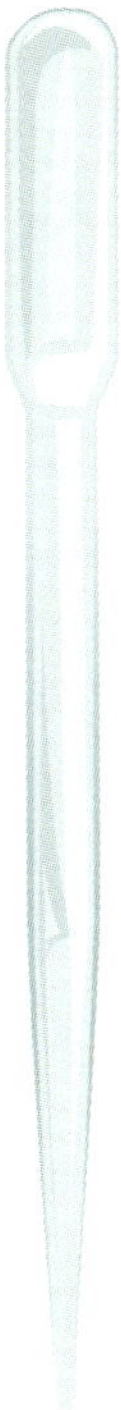

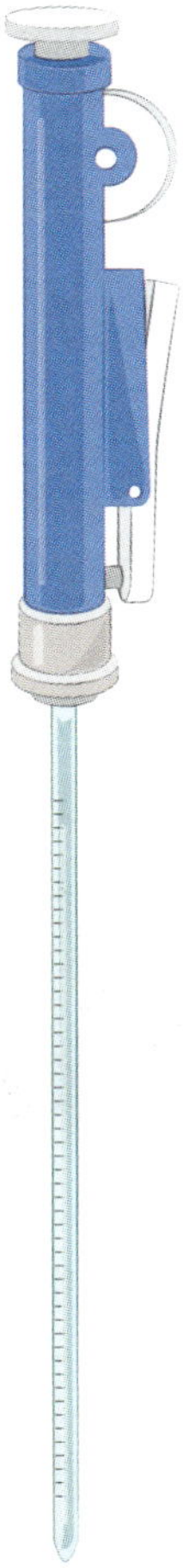

pipette

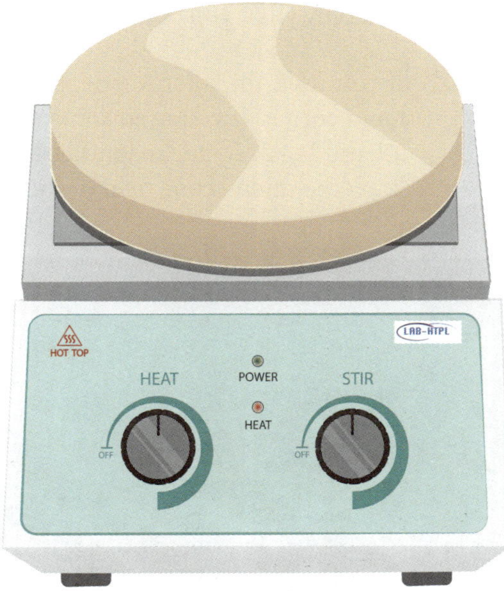

hot plate

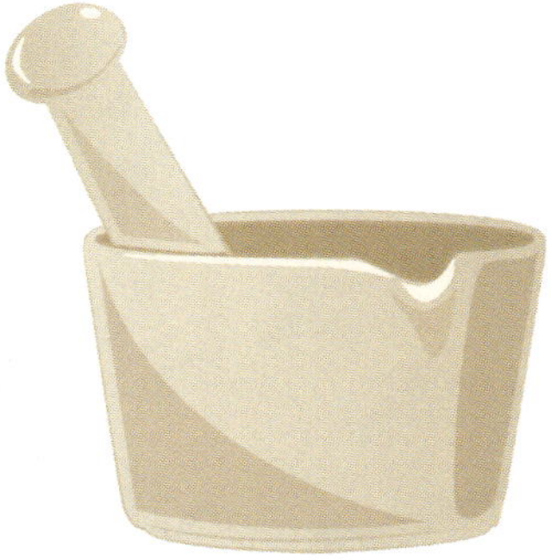

mortar and pestle

Appendix C

MAKING BIOLOGICAL DRAWINGS

During some lab activities, you will be asked to draw what you observe. Drawing scientific specimens is one of the best ways to learn some of the complex biological structures and processes that you will be observing. Drawing is also a great field skill to have. As you draw, concentrate on the shape and textures of the different structures you observe. By the time you have finished, you should better understand the structures that you have drawn.

1. Write the name of the specimen above the drawing. If the drawing is of a portion of an organism, indicate the portion in the title or under the title. For example, the title "Fruit Fly" alone indicates a drawing of the entire fly, while "Fruit Fly, leg" indicates that the drawing is of the leg of a fruit fly.

2. If the specimen or structure has been prepared as a wet mount or a longitudinal mount before you observe it, indicate this under the name of the specimen.

3. Make the drawing large and center it in the space available.

4. You can use just a pencil for most drawings so that you can erase anything you want to change. If it makes your drawing clearer, use colored pencils, though a colorful drawing should not be a goal in itself.

5. Add labels after the drawing is complete; label what you can identify and only what you observe.

6. If you used the microscope, indicate the power in the lower right-hand corner. If you used some other type of magnification (such as a hand lens), write "magnified" or "enlarged" in that corner. Don't draw the microscope field or the container that the specimen is in.

HOW TO PREPARE A WET MOUNT

When you use a microscope to look at specimens, sometimes you will use slides that have been prepared ahead of time. For some lab activities, however, you will need to make your own temporary slide. Placing a specimen in a drop of water to observe it is called making a *wet mount*. To make a wet mount, you need a microscope slide, a cover slip, and the specimen that you want to observe.

1. Handle glass slides and cover slips by their edges so that you do not leave fingerprints on them (as shown below).

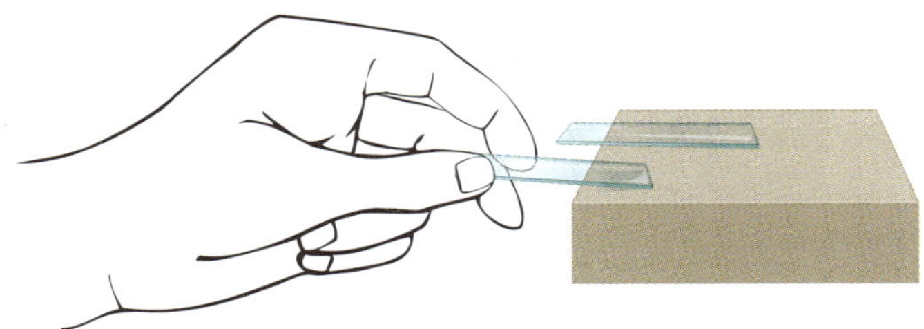

2. Be careful not to bend cover slips. If you are using a glass cover slip, handle it gently. Splinters from shattered glass cover slips easily enter the fingers and may require surgical removal. Inspect plastic cover slips for excessive scratches. If too many scratches appear, discard the cover slip.

3. Use only water when washing the slides and cover slips. Soap film may kill or damage living specimens. Shake off excess water, then dry the slide with a tissue.

4. Place a drop of water in the center of the slide. Using forceps, place the specimen in the drop of water. In some cases, your teacher may provide a special concavity slide that has a shallow well to hold larger specimens.

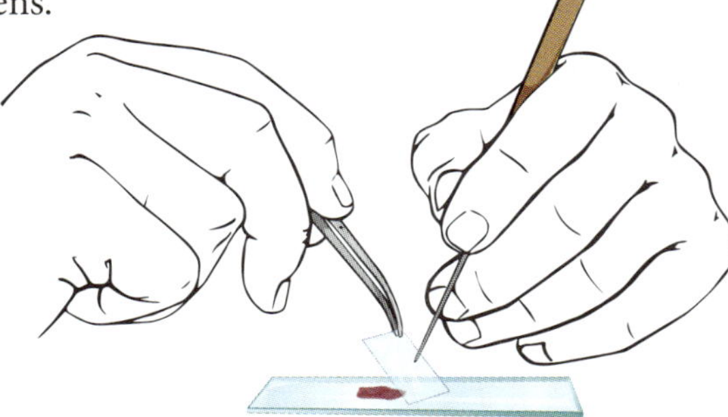

5. Place the cover slip on top of the specimen so that one edge is touching the slide and the cover slip is held at a 45° angle above the drop of water. Slowly lower the cover slip down on top of the water and specimen. If bubbles appear in the area that you are going to view, tap the cover slip with the tip of a probe to remove them.

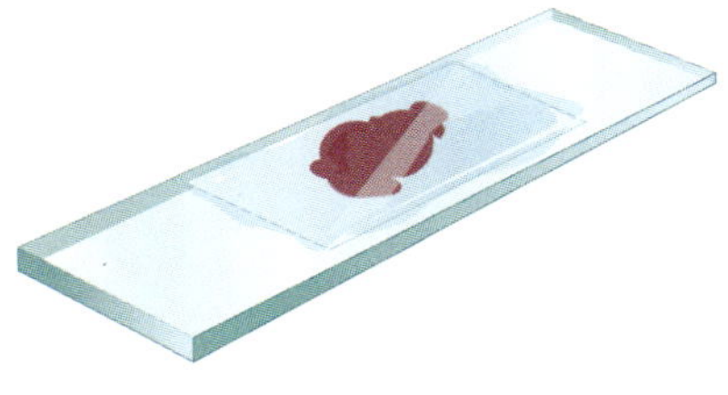

6. When finished observing, dispose of the specimen and wash the slide and cover slip, placing them on a paper towel to air dry.

Appendix C

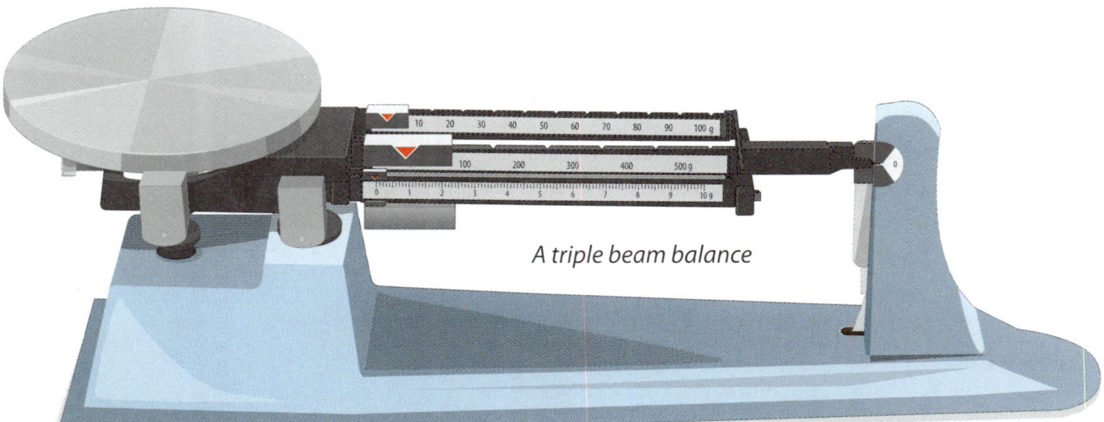

A triple beam balance

USING A MECHANICAL BALANCE

The mass of a substance can be determined in the laboratory with the use of a mechanical balance. Several kinds of mechanical balances are common, but all of them operate on the same principles. To use a mechanical balance properly, follow the steps given below.

1. Place the balance on a smooth, level surface.

2. Keep the balance pan(s) clean and dry. Never put chemicals directly on the metal surface of the pan(s). Place materials on a sheet of weighing paper, filter paper, watch glass, in a weighing boat, or in a beaker.

3. Check the rest point of the empty balance. To do this, remove all weight from the pans and slide all movable masses to their zero positions. If the balance beam swings back and forth, note the central point of the swing. You do not have to wait until the beam stops swinging completely. If the central point on the balance arm does not align with the marked zero point on the post, ask your teacher to adjust the balance. *Do not adjust the balance yourself!*

4. Place the substance to be measured on the pan and adjust the sliding masses. Move the largest mass first, and then make final adjustments with the smaller masses. The sum of all the readings is the mass of the object (as show below). Be sure to subtract the mass of the container or paper on which you placed the substance.

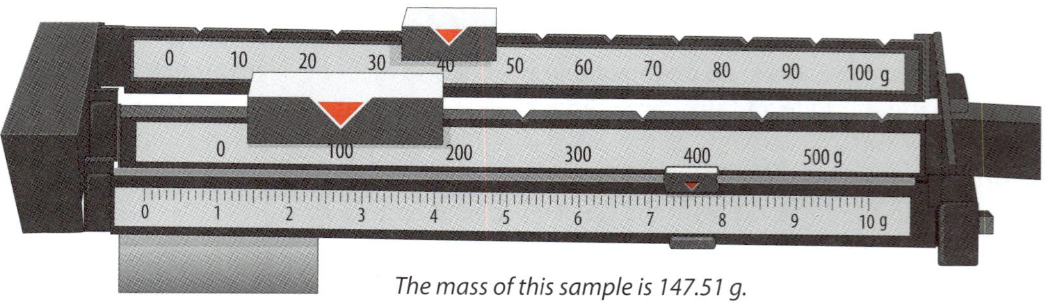

The mass of this sample is 147.51 g.

USING AN ELECTRONIC BALANCE

Electronic balances are generally faster and easier to use than their mechanical counterparts. To use an electronic balance properly, follow the instructions given below.

1. Place the balance on a smooth, level surface.

2. Keep the balance pan(s) clean and dry. Never put chemicals directly on the metal surface of the pan(s). Place materials on a sheet of weighing paper, filter paper, watch glass, in a weighing boat, or in a beaker.

3. Turn the balance on. Place the container or weighing paper that will hold the substance to be measured and make sure that there is a reading of 0 by pushing the **Tare** button.

4. Place your substance on the paper or in the container and read the mass in the display. If measuring out a predetermined amount of a substance, add the substance until you have reached the appropriate mass.

USING A THERMOMETER

When using a thermometer in lab activities, make sure that you are using one that has the proper temperature range for the experiment that you will be doing. Always hold a thermometer in the container. *Never leave a thermometer standing upright in a container; the container is likely to get knocked over.*

Position the thermometer bulb just above the bottom of the container. If the bulb touches the container, your readings will be inaccurate.

If a thermometer breaks, alert your teacher and do not touch the inner contents. Some thermometers contain mercury. The spilled mercury may look fascinating, but it is toxic and can be absorbed through the skin.

HANDLING LIQUIDS

Proper technique for handling liquids is essential if you are to remain safe, keep reagents pure, and obtain accurate measurements. For increased safety, do not splash or splatter liquids when pouring. Pour them slowly down the insides of test tubes and beakers. If anything is spilled, clean it up quickly. (See First Aid Rules on page 250.)

In order to keep the liquid chemicals pure, keep stirring rods out of the stock supply. Do not let the stoppers and lids become contaminated while you are pouring. Instead, hold the stopper between your fingers. If you must put a lid down, keep the inside surface from touching the surface of the table (as shown at right).

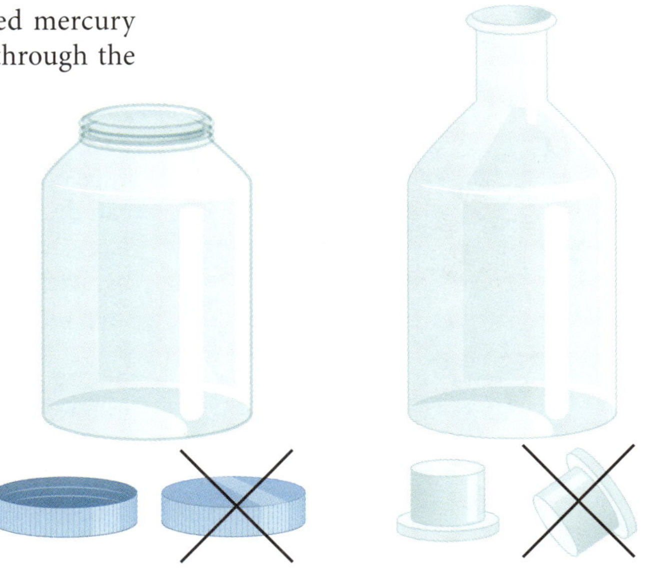

Don't let the part of a lid that could touch a chemical also touch the table.

Appendix C

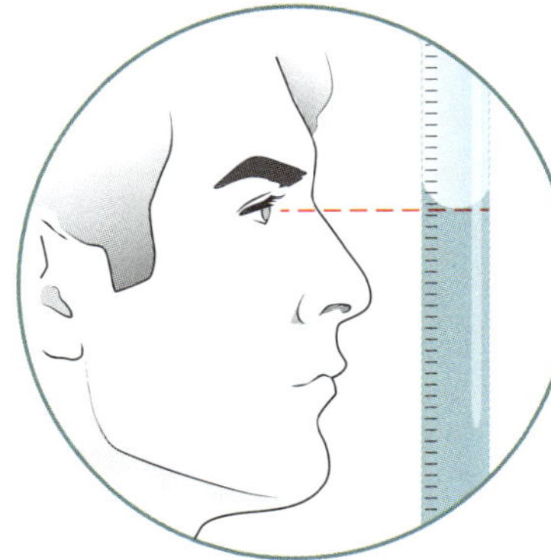

Accurate measurements of liquids can be made in burets, graduated cylinders, and volumetric flasks. You should measure volumes in these pieces of glassware unless you need only a rough approximation. When reading the level of a liquid, look at the bottom of the meniscus (curved surface) along a horizontal line of sight (as shown at left).

Measuring out solids: (1) Scoop out a little of the sample with the spatula. (2) Gently tap the spatula until the desired amount falls off onto the paper. (3) Cup the paper to pour a powdered solid into a test tube.

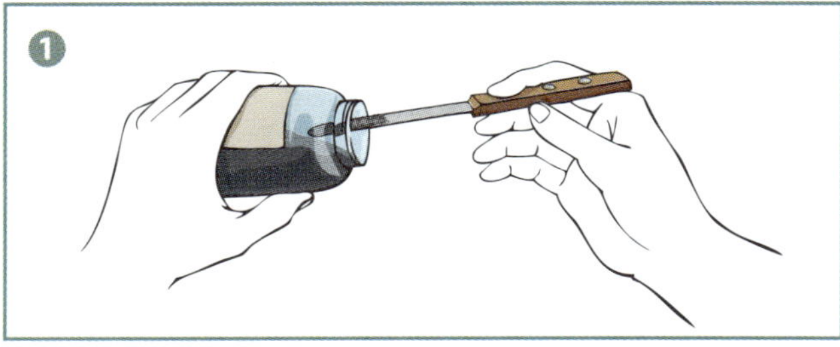

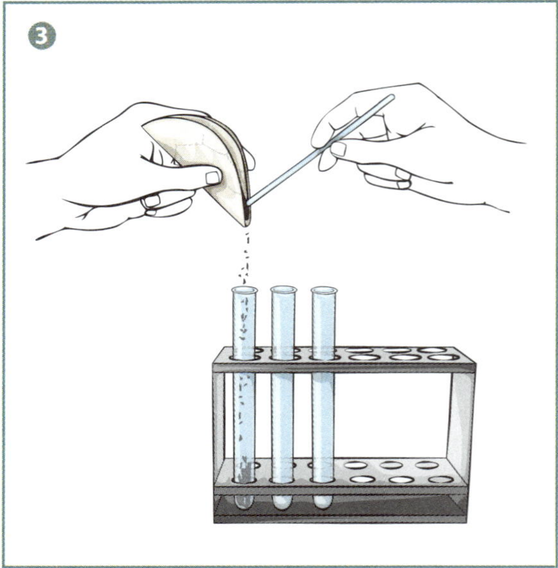

METHODS OF SLOWING PROTOZOAN MOVEMENT

In some lab activities you may be working with protozoans, microscopic organisms that move. This makes them hard to observe! You can slow down protozoans and make them easier to observe in a wet mount by using a cover slip, cotton fibers, or a thicker medium.

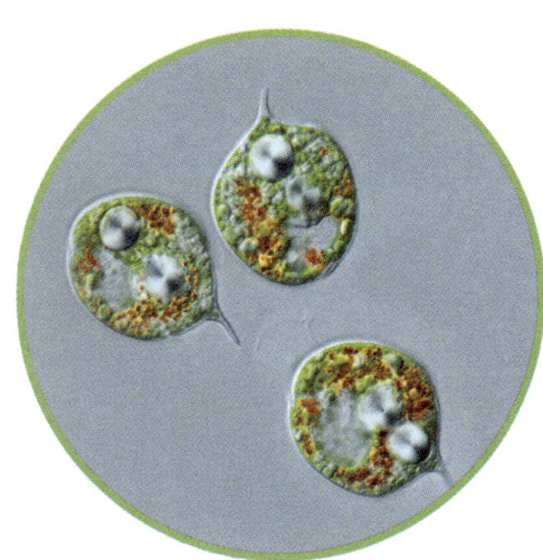

As the culture medium of a wet mount evaporates, the cover slip will press on the organism and slow its movement. A paper towel on the edge of the cover slip can be used to speed the process. Your lab partner can draw off small portions of the medium while you chase the organism. Do not permit the medium to evaporate completely. Replenish it by placing a drop of medium beside the cover slip and letting some of it seep under the cover slip.

Before you place the cover slip on the medium containing protozoans in a wet mount, you can place a small quantity of cotton fibers on the medium. These serve as obstacles, blocking the path of protozoans and thus localizing their activities.

Special media (glycerin, methyl cellulose, or commercially prepared products) can be used to slow protozoans. Because these media are thicker than water, the protozoans move more slowly through them.

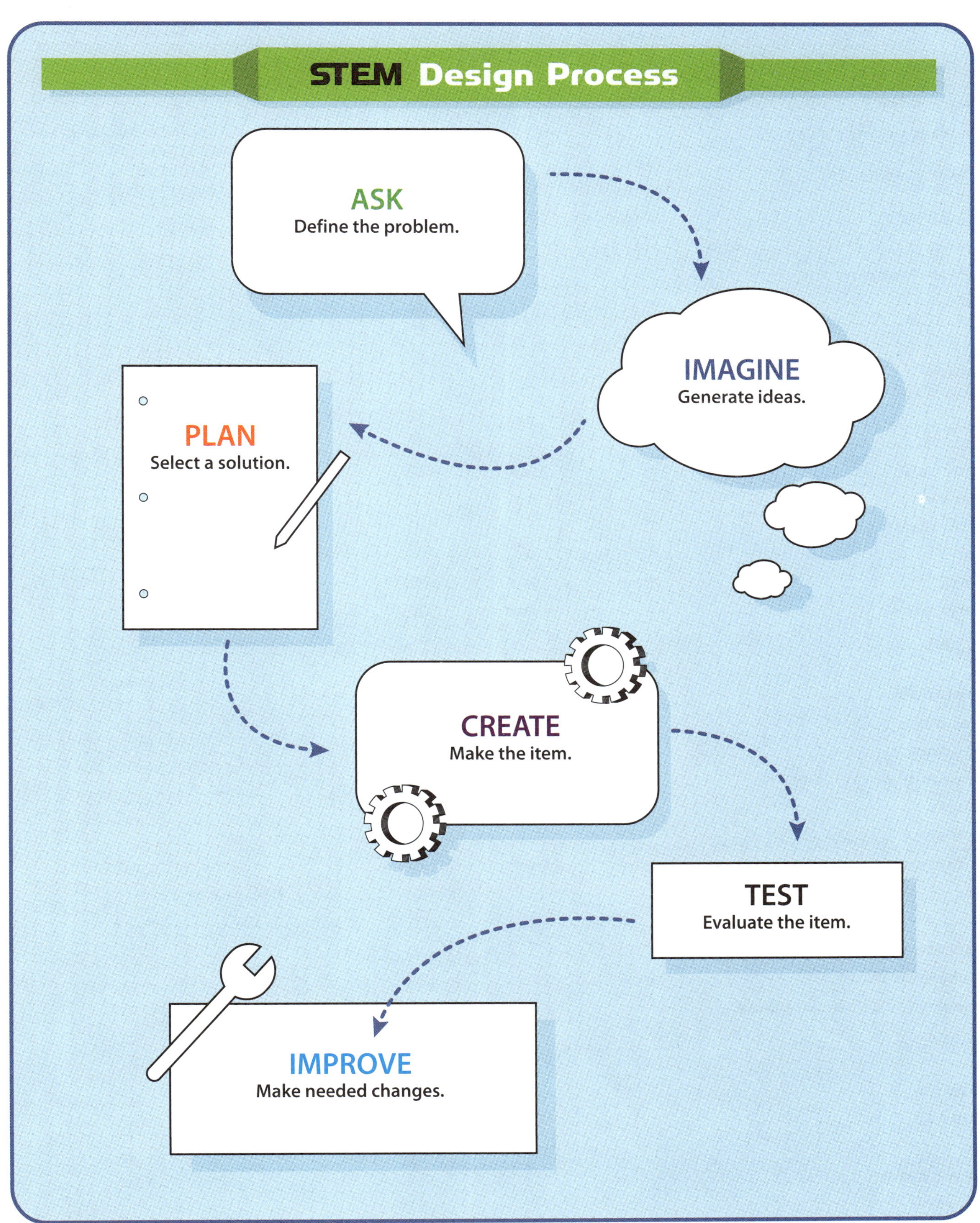
STEM Design Process
ASK
Define the problem.
IMAGINE
Generate ideas.
PLAN
Select a solution.
CREATE
Make the item.
TEST
Evaluate the item.
IMPROVE
Make needed changes.

Equipment and Materials List (alphabetical)

Amount required is maximum for each lab group in any one lab

Item	Size	# Req.	Lab	Remarks
alcohol	100 mL	1	2E	70% ethanol or 70% isopropyl
aluminum foil	8 cm × 8 cm	3	2E	
amoebas, culture of living		1	7C	available at most science supply companies
amylase solution, 2%	20 mL		14E	
	48 mL		6D	
apple juice	200 mL		7D	
aquarium	small	1	11E	
balance, laboratory		1	8D	
balloon		4	7D	
beaker	250 mL	2	2D	
	400 mL	1	2E	
	600 mL	1	14E	
		3	7D	large enough to hold the bottles
Benedict's solution	25 mL		14E	
binoculars		1 pair	18C	optional
blender		1	20D	
BMR calculator, online			14D	
board	1 in. × 4 in. × 10 ft	1	20D	to build frame
bottle	20 oz	4	7D	
bottle, plastic	varied	several	12D	e.g., water, soda, milk
bowl		3	9D	
		1	14E	
bowl, shallow		6	8D	
box, empty		1	4E	
calculator		1	6C	
Calories calculator, online			14D	
camera		1	10E	
cloth piece			20D	substitute: blotting paper
cloth square, thin	90–120 cm	several	12D	e.g., tulle, muslin, or cotton bed sheets
coin		3	4D	penny, nickel, quarter (1 each)
computer with Excel spreadsheet		1	6C, 12C, 17E	substitute: any spreadsheet program
computer with Internet access		1	17D	
container, plastic	varied	several	12D	e.g., yogurt, sour cream
container, mug or drinking glass		3	11D	Each needs a lid with a hole in it.
cotton ball		8	9C	
		1 bag	12D	
cover slip		several	1D, 7C	
craft stick		1 pack	12D	
cup		several	12D	plastic, various colors
cutting board		1	8D	
die, game		1	4E	

Item	Size	# Req.	Lab	Remarks
dishpan	large	1	20D	must be larger than 11 in. × 18 in.
dowel, wooden	30–60 cm	several	12D	substitute: bamboo sticks
dumbbell		1	13F	substitute: any similar weight
eyedropper		1	9C	
feathers, down	100 g		11D	
fertilizer solution	500 mL		9D	Follow directions on the fertilizer container.
field guide		several	10E, 19E	substitute: keys
field notebook		1	18C	
fish, live		1	11E	
flowerpot		5	5D	
foil, aluminum		1 box	12D	
food coloring			2D	dark color
freezer		1	11D	
fresh specimens, slide of		various	1D	microorganisms, optional
glass (drinking)	large	2	16D	substitute: cups
glass plate		2	9C	
glue			12D	
goggles		1 pair	2D, 2E, 6D, 14E	
graduated cylinder	10 mL	1	6D, 14E	
graduated cylinder	50 mL	1	7D	
hand lens		1	18C	optional
heavy object		1	20D	e.g., large book
hot plate		1	2E	substitute: any heat source without an open flame
hot plate		1	6D	not needed if your water heater can produce 37 °C water
hot plate		1	14E	
ice	1 small bag		7D	
ice cubes		several	11E	
iodine solution	10 mL	1	2E	
iodine solution			6D	
jar	various	several	12D	glass or plastic
knife		1	8D	
label, paper		6	9D	
LED puck light		several	12D	
marbles, colored		60	4E	10 each of 6 different colors
marshmallows, mini, colored	miniature	2	4D	green
marshmallows, mini, colored	miniature	24	4D	pink, white, yellow (8 each color)
metal can		several	12D	e.g., coffee, peanut
microscope		1	1D, 3D, 7C, 13E	
mortar and pestle		1	14E	
newspaper spread		2	20D	
notebook, small		1	10E	
nylon screen square	30–60 cm	several	12D	

Item	Size	# Req.	Lab	Remarks
paper		1 pack	12D	various colors
paper clip		6	2E	
paper towel		several	9C	
paper, white		2 sheets	16E	
pencil, marking		1	9C	
		1	14E	
petri dish		1	2E	
pipe cleaner		3	4D	
pipette, disposable		31	6D	
		several	7C	
pitch app		1	17E	
plant, geranium or coleus		1	2E	
plant specimen	various		8C	
plastic bag		1	11E	
plate		several	12D	plastic, various colors
polyethylene glycol solution, 20% w/v	500 mL		9D	Dissolve 100 g of PEG in 250 mL of water and then dilute the solution with enough water to make 500 mL of solution.
popped popcorn		2	4D	substitute: puffed-corn cereal
potato		1	4D	
potato, fresh		1	8D	
potting soil			5D, 9D	One large bag purchased at the beginning of the year should be sufficient for both labs.
propylene glycol			12D	Do not use ethylene glycol because of its danger to mammals.
pulse monitor		1	15E	substitute: pulse monitor app
pushpin or thumbtack, colored		8	4D	green, red, white, yellow (2 each)
Q-tip		1 bag	12D	
random letter generator			5E	commonly available online
random number generator			5E, 6D, 16D	commonly available online
reference resources			12C	
refrigerator		1	9D	
respiration rate app		1	15D	substitute: stopwatch
rubber band		several	9C	
running app		1	14D	substitute: pedometer
saltine crackers	pack	1	14E	
salt water, saturated			8D	Quantity will vary depending on size of potato strips and containers used.
sandpaper		1 piece	9D	
scissors		several	12D	
screws	1 1/4 in.		20D	to build frame
seed starting tray	72-cell	1	9D	Use 1 tray for the entire class.
seeds, bean		8	9C	
seeds, irradiated	pack	4	5D	

Item	Size	# Req.	Lab	Remarks
seeds, nonirradiated	pack	1	5D	
seeds, watermelon		12	9D	
shovel, hand		1	12D	
slide, concavity		several	7C	
slide, microscope		various	1D	blank
slide, prepared		various	1D	of various kinds of cells
		various	3D	of onion, hyacinth, or lily root tips showing phases of plant mitosis
		several	13E	of human skin
sponge	large	1	20D	
spoon, measuring		2	2D	
staple gun with staples		1	20D	
starch solution, 2%	30 mL		6D	
stirring rod		1	2D	substitute: spoon
stopwatch		1	2D, 7D, 13F, 14D, 15E, 16E	substitute: clock with second hand
straw, drinking		1 pack	12D	
string			12D	
sugary liquid			12D	e.g., sugar water, soda, juice
tape			12D	
test tube		30	6D	
		5	14E	
test tube rack		3	6D	
		1	14E	
test tube tongs		1	14E	
thermometer		4	7D	
		3	11D	
		1	11E	substitute: temperature probe
		2	16D	
thermometer, strip		2	13F	
toothpick		16	4D	
twist tie		multiple	12D	
water			2E, 8D, 9C, 9D, 14E, 20D	
water, 37 °C			6D	
water, cold			16D	
water, cold and hot			2D	
water, hot			11E, 16D	
water, warm			7D	
window screen	16 × 13 in.	1	20D	to build frame
wire, floral			12D	
wool, unspun	600 g		11D	
yeast	20 g		7D	

Equipment and Materials List (by lab)

Amount required is specified for each lab group

Lab	Item	Size	# Req.	Remarks
1D	cover slip		various	
	fresh specimens, slide of		various	microorganisms, optional
	microscope		1	
	slide, microscope		various	blank
	slide, prepared		various	of various kinds of cells
1E	none			
2D	beaker	250 mL	2	
	food coloring			dark color
	goggles		1 pair	
	spoon, measuring		2	
	stirring rod		1	substitute: spoon
	stopwatch		1	substitute: clock with a second hand
	water			hot and cold
2E	alcohol	100 mL	1	70% ethanol or 70% isopropyl
	aluminum foil	8 cm × 8 cm	3	
	beaker	400 mL	1	
	goggles		1 pair	
	hot plate		1	substitute: any heat source without an open flame
	iodine solution	10 mL	1	
	paper clip		6	
	petri dish		1	
	plant, geranium or coleus		1	
	water			
3D	microscope		1	
	prepared slides		various	of onion, hyacinth, or lily root tips showing phases of plant mitosis
4D	coin		3	penny, nickel, quarter (1 each)
	marshmallows, mini, colored	miniature	2	green
	marshmallows, mini, colored	miniature	24	pink, white, yellow (8 each color)
	pipe cleaner		3	
	popped popcorn		2	substitute: puffed-corn cereal
	potato		1	
	pushpin or thumbtack, colored		8	green, red, white, yellow (2 each)
	toothpick		16	
4E	box, empty		1	
	die, game		1	
	marbles, colored		60	10 each of 6 different colors
5D	flowerpot		5	
	potting soil			One large bag purchased at the beginning of the year should be sufficient for Labs 5D and 9D.
	seeds, irradiated	pack	4	
	seeds, nonirradiated	pack	1	

Lab	Item	Size	# Req.	Remarks
5E	random letter generator			commonly available online
	random number generator			commonly available online
6C	calculator		1	
	computer with Excel spreadsheet		1	substitute: any spreadsheet program
6D	amylase solution, 2%	48 mL		
	goggles		1 pair	
	graduated cylinder	10 mL	1	
	hot plate		1	not needed if your water heater can produce 37 °C water
	iodine, tincture			
	pipette, disposable		31	
	random number generator			commonly available online
	starch solution, 2%	30 mL		
	test tube		30	
	test tube rack		3	
	water			37 °C
7C	amoebas, culture of living		1	available at most science supply companies
	cover slip		several	
	microscope		1	
	pipette, disposable		several	
	slide, concavity		several	
7D	apple juice	200 mL		
	balloon		4	
	beaker		3	large enough to hold the bottles
	bottle	20 oz	4	
	graduated cylinder	50 mL	1	
	ice	1 small bag		
	stopwatch		1	
	thermometer		4	
	water			warm
	yeast	20 g		
8C	plant specimen	various	1	
8D	balance, laboratory		1	
	bowl, shallow		6	
	cutting board		1	
	knife		1	
	potato, fresh		1	
	salt water, saturated			Quantity will vary depending on size of potato strips and containers used.
	water			
9C	cotton ball		8	
	eyedropper		1	
	glass plate		2	

Equipment and Materials List (by lab)

Lab	Item	Size	# Req.	Remarks
9C	paper towel		several	
	pencil, marking		1	
	rubber band		several	
	seeds, bean		8	
	water			
9D	bowl		3	
	fertilizer solution	500 mL		Follow directions on the fertilizer container.
	labels, paper		6	
	polyethylene glycol solution, 20% w/v	500 mL		Dissolve 100 g of PEG in 250 mL of water and then dilute the solution with enough water to make 500 mL of solution.
	potting soil			See Lab 5D.
	refrigerator		1	
	sandpaper		1 piece	
	seed starting tray	72-cell	1	Use 1 tray for the entire class.
	seeds, watermelon		12	
	water			
10D	none			
10E	camera		1	
	field guide		several	
	notebook, small		1	
11D	container, mug or drinking glass		3	Each needs a lid with a hole in it.
	feathers, down	100 g		
	freezer		1	
	thermometer		3	
	wool, unspun	600 g		
11E	aquarium	small	1	
	fish, live		1	
	ice cubes		several	
	plastic bag		1	
	thermometer		1	substitute: temperature probe
	water, hot			
12C	computer with Excel spreadsheet		1	substitute: any spreadsheet program
	reference resources			
12D	bottle, plastic		several	e.g., water, soda, milk
	cloth square, thin	90–120 cm	several	e.g., tulle, muslin, or cotton bed sheets
	container, plastic		several	e.g., yogurt, sour cream
	cotton ball		1 bag	
	craft stick		1 pack	
	cup		several	plastic, various colors
	dowel, wooden	30–60 cm	several	substitute: bamboo sticks
	foil, aluminum		1 box	
	glue			

Lab	Item	Size	# Req.	Remarks
12D	jar	various	several	glass or plastic
	LED puck light		several	
	metal can		several	e.g., coffee, peanut
	nylon screen square	30–60 cm	several	
	paper		1 pack	various colors
	plate		several	plastic, various colors
	propylene glycol			Do not use ethylene glycol because of its danger to mammals.
	Q-tip		1 bag	
	scissors		several	
	shovel, hand		1	
	straw, drinking		1 pack	
	string			
	sugary liquid			e.g., sugar water, soda, juice
	tape			
	twist tie		multiple	
	wire, floral			
13E	microscope		1	
	prepared slides		several	of human skin
13F	dumbbell		1	substitute: any similar weight
	stopwatch		1	substitute: clock with second hand
	thermometer, strip		2	
14D	BMR calculator, online			
	Calories calculator, online			
	running app		1	substitute: pedometer
	stopwatch		1	
14E	amylase solution, 2%	20 mL		
	beaker	600 mL	1	
	Benedict's solution	25 mL		
	bowl		1	
	goggles		1 pair	
	graduated cylinder	10 mL	1	
	hot plate		1	
	mortar and pestle		1	
	pencil, marking		1	
	saltine crackers	pack	1	
	test tube		5	
	test tube rack		1	
	test tube tongs		1	
	water			
15D	respiration rate app		1	substitute: stopwatch
15E	pulse monitor		1	substitute: pulse monitor app
	stopwatch		1	

Equipment and Materials List (by lab)

Lab	Item	Size	# Req.	Remarks
16D	glass (drinking)	large	2	substitute: cups
	random number generator			commonly available online
	thermometer		2	
	water, cold			
	water, hot			
16E	paper, white		2 sheets	
	stopwatch		1	
17D	computer with Internet access		1	
17E	computer with Excel spreadsheet		1	substitute: any spreadsheet program
	pitch app		1	
18C	binoculars		1 pair	optional
	field notebook		1	
	hand lens		1	optional
19D	none			
19E	field guide or key			several
20D	blender		1	
	board	1 in. × 4 in. × 10 ft	1	to build frame
	cloth piece			substitute: blotting paper
	dishpan	large	1	must be larger than 11 in. × 18 in.
	heavy object		1	e.g., large book
	newspaper spread		2	
	screws	1 1/4 in.		to build frame
	sponge	large	1	
	staple gun with staples		1	
	water			
	window screen	16 × 13 in.	1	to build frame
20E	none			

Photo Credits

Key: (t) top; (c) center; (b) bottom; (l) left; (r) right; (bg) background; (i) inset

Cover

front Tirik/Shutterstock.com; **spine** IrinaK/Shutterstock.com; **back** © iStock.com/photomaru

Front Matter

i bogdanhoda/Shutterstock.com; **iii** Anton Nalivayko /Shutterstock.com; **iv** © iStock.com/Salima Senyavskaya; **v** Sari ONeal/Shutterstock.com; **vi** Top Photo Engineer/Shutterstock .com; **vii** Africa Studio/Shutterstock.com; **viiit** © iStock.com /mediaphotos; **viiib** korkeng/Shutterstock.com; **ix** Marek Mis /Science Source; **xt** Barry Barnes/Shutterstock.com; **xb** Gino Santa Maria/Shutterstock.com; **xi** © iStock.com/BanksPhotos; **xii** Oleksandr Kostiuchenko/Shutterstock.com; **xiii** Air Images /Shutterstock.com

Chapter 1

1 KENG MERRY Paper Art/Shutterstock.com; **3** Photok.dk /Shutterstock.com; **5** amenic181/Shutterstock.com; **6** (petri dish) Jarun Ontakra/Shutterstock.com; **6–7** (popcorn) andersphoto /Shutterstock.com; **7t** Evgeny Parushin/Shutterstock.com; **7b** Melinda Fawver/Shutterstock.com; **8** KellyNelson/Shutterstock .com; **9** 279photo Studio/Shutterstock.com; **10, 11r** Triff /Shutterstock.com; **11l** Natalia_Arefieva/Shutterstock.com; **12** Wave_Movies/Shutterstock.com; **14** Phil Boorman/Cultura/Getty Images; **15tl** SomeSense/Shutterstock.com; **15tr** Jeff Hutchens /Getty Images News/Getty Images; **15cl** jiangdi/Shutterstock.com; **15cr** Jesser Chio/EyeEm/Getty Images; **15bl** © iStock.com /epantha; **15br** © iStock.com/Max_Xie; **15–17bg** korkeng /Shutterstock.com; **16** (tapeworm) James H. Robinson/Science Source; **16** (millipede) asawinimages/Shutterstock.com; **16** (octopus) zhengzaishuru/Shutterstock.com; **16** (hawk) Stephen Mcsweeny/Shutterstock.com; **16** (tick) Allahfoto/Shutterstock .com; **16** (urchin) eye-blink/Shutterstock.com; **16** (crayfish) Zheltyshev/Shutterstock.com; **16** (snail) © iStock.com/aristotoo; **16** (leech) xpixel/Shutterstock.com; **16** (bat) Rosa Jay /Shutterstock.com

Chapter 2

19tl CNRI/Science Source; **19cl** Steeve Roche/Shutterstock.com; **19bl** © iStock.com/WhitcombeRD; **19tr** Steve Gschmeissner /Science Source; **19cr** Biophoto Associates/Science Source; **19br** science photo/Shutterstock.com; **21** Ola Ko/Shutterstock.com; **23** Pekka Nikonen/Shutterstock.com; **25** Africa Studio/Shutterstock .com; **26–27t, 28** Roman Sigaev/Shutterstock.com; **26–27b** Turtle Rock Scientific/Science Source; **29** BJU Press; **30** Cordelia Molloy/ Science Source; **31** Elpis Ioannidis/Shutterstock.com

Chapter 3

35 eranicle/Shutterstock.com; **39l** Scenics & Science/Alamy Stock Photo; **39r, 40i** Claudio Divizia/Shutterstock.com; **40** © iStock .com/Lena_Zajchikova

Chapter 4

41 Lindsay Keats/Alamy Stock Photo; **42** Juniors Bildarchiv GmbH/Alamy Stock Photo; **43** Ramon L. Farinos/Shutterstock .com; **44** © iStock.com/stocknroll; **45l, 46** @Hans Surfer/Moment /Getty Images; **45r** Dorottya Mathe/Shutterstock.com; **47** BJU Photo Services; **48–49** (pins) Faizal Ramli/Shutterstock.com; **48** (popcorn) Jiri Hera/Shutterstock.com; **48** (potatoes) Jacek Fulawka/Shutterstock.com; **48** (toothpicks) panuwat panyacharoen/Shutterstock.com; **49** bigacis/Shutterstock.com; **51** KENG MERRY Paper Art/Shutterstock.com; **52** Gary Paul Lewis /Shutterstock.com; **53** Numpon Jumroonsiri/Shutterstock.com

Chapter 5

55 dimair/Shutterstock.com; **56** Herschel Hoffmeyer /Shutterstock.com; **57** leungchopan/Shutterstock.com; **59** tasani bin abdul hamid/Shutterstock.com; **61** Egorov Igor/Shutterstock .com; **62** Alex_Po/Shutterstock.com; **63** Szasz-Fabian Jozsef /Shutterstock.com; **65** Mopic/Shutterstock.com; **67** andriano.cz /Shutterstock.com

Chapter 6

69 Pierre Jean Durieu/Shutterstock.com; **73** Per Bengtsson /Shutterstock.com; **75** © iStock.com/iLexx; **77** Stanislav Salamanov/Shutterstock.com; **78** Andrew Lambert Photography /Science Source / Africa Studio/Shutterstock.com; **79t** Ilike /Shutterstock.com; **79b** Xinhua/Alamy Stock Photo

Chapter 7

81t Darkfoxelixir/Shutterstock.com; **81c** STEVE GSCHMEISSNER/SCIENCE PHOTO LIBRARY/Getty Images; **81b** Stocktrek Images/Media Bakery; **82t** Henri Koskinen /Shutterstock.com; **82ct** Jubal Harshaw/Shutterstock.com; **82cb** Flip Nicklin/Minden Pictures/Getty Images; **82b** Eric V. Grave /Science Source; **83** Pieter Bruin/Shutterstock.com; **85** sergiophoto/Shutterstock.com; **86bg** Ksw Photographer /Shutterstock.com; **86** Steve Gschmeissner/Science Source; **87l** xpixel/Shutterstock.com; **87c** MaraZe/Shutterstock.com; **87r** exopixel/Shutterstock.com; **88t** Christos Theologou/Shutterstock .com; **88b** JIANG HONGYAN/Shutterstock.com

Chapter 8

91l vladwel/Shutterstock.com; **91r** Pekka Nikonen/Shutterstock .com; **91ri** Glasscage/Shutterstock.com; **93t** Organica/Alamy Stock Photo; **93b** © iStock.com/YinYang; **95l** Picsfive /Shutterstock.com; **95r** photka/Shutterstock.com; **96, 97** (potatoes) Jacek Fulawka/Shutterstock.com; **97** Richard Bradford /Shutterstock.com

Chapter 9

99l kongsky/Shutterstock.com; **99r** Ammak/Shutterstock.com; **100t** Alena Brozova/Shutterstock.com; **100b** © iStock.com /Jolkesky; **101t** Ben Schonewille/Shutterstock.com; **101b** Africa Studio/Shutterstock.com; **103** Martin Shields/Alamy Stock Photo; **104** D_M/Shutterstock.com; **105** Melica/Shutterstock.com; **107** PORNPIPAT CHAROENTHAI/Shutterstock.com; **108** Olga Ilina/ Shutterstock.com; **109** Indigo Photo Club/Shutterstock.com

Chapter 10

111tr saad315/Shutterstock.com; **111l** Tim Painter/EyeEm/Getty Images; **111br** William D. Bachman/Science Source; **112t** © iStock.com/Omenn; **112b** Josef Pittner/Shutterstock.com; **113** christos Theologou/Shutterstock.com; **115** Jim Cumming /Shutterstock.com; **117** zhekoss/Shutterstock.com; **119t** Eco/UIG /Universal Images Group/Getty Images; **119b** Eric Isselee

/Shutterstock.com; **121–23**bg korkeng/Shutterstock.com; **121**t Dawna Moore/Alamy Stock Photo; **121**b © iStock.com/arlindo71; **122**t Andi111/Shutterstock.com; **122**c Doug McLean /Shutterstock.com; **122**b Tony Campbell/Shutterstock.com; **123** GCapture/NumbSt/Shutterstock.com

Chapter 11

129tl alslutsky/Shutterstock.com; **129**tc itor/Shutterstock.com; **129**tr, **131**r Eric Isselee/Shutterstock.com; **129**bl © iStock.com /Antagain; **129**br © iStock.com/hdere; **131**l DenisNata /Shutterstock.com; **132** (jars) donatas1205/Shutterstock.com; **132** (thermometer) Tarzhanova/Shutterstock.com; **132**b (feathers) Palmer Kane LLC/Shutterstock.com; **132**t (wool) Melica /Shutterstock.com; **133** Pete Pahham/Shutterstock.com; **135** bogonet/Shutterstock.com; **136–37**t Good Shop Background /Shutterstock.com; **136**b Mikael Damkier/Shutterstock.com; **137**b, **138**b Alexandra Lande/Shutterstock.com; **137**c Vangert /Shutterstock.com; **138**t Africa Studio/Shutterstock.com

Chapter 12

139 Kucher Serhii/Shutterstock.com; **141** ESB Professional /Shutterstock.com; **143** LMLPhoto/Shutterstock.com; **144**t Olga Zarytska/Shutterstock.com; **144**b S-F/Shutterstock.com; **145** anyaivanova/Shutterstock.com; **147**t alle/Shutterstock.com; **147**bl Rudmer Zwerver/Shutterstock.com; **147**br nico99/Shutterstock .com; **148** ILYA AKINSHIN/Shutterstock.com; **149** Mr. SUTTIPON YAKHAM/Shutterstock.com; **150** Kim Taylor and Jane Burton/Dorling Kindersley/Getty Images

Chapter 13

151 LightField Studios/Shutterstock.com; **153** Anton Nalivayko /Shutterstock.com; **155**r Sebastian Kaulitzki/Shutterstock.com; **155**l © iStock.com/Eraxion; **157**t Monkey Business Images /Shutterstock.com; **157**ti Ramona Kaulitzki/Shutterstock.com; **157**b © Martinmark | Dreamstime.com; **157**bi Ramona Kaulitzki /Shutterstock.com / SCIEPRO/Science Photo Library/Getty Images; **159** nelen/Shutterstock.com; **160** Jose Luis Calvo /Shutterstock; **161** exopixel/Shutterstock.com; **163** karelnoppe /Shutterstock.com; **164** wheatley/Shutterstock.com; **165** oliveromg/Shutterstock.com

Chapter 14

167 artjazz/Shutterstock.com; **168** Bongarts/Getty Images; **169**, **171**l Chipmunk131/Shutterstock.com; **169**i Vecton/Shutterstock .com; **171**li © iStock.com/ikuvshinov; **171**r Designua /Shutterstock.com; **173** rawiwano/Shutterstock.com; **174** Maridav/Shutterstock.com; **175** © iStock.com/Sanny11; **176** Cozine/Shutterstock.com; **177** michaeljung/Shutterstock.com

Chapter 15

179l Chipmunk131/Shutterstock.com; **179**li, **183** Vecton /Shutterstock.com; **179**r BlueRingMedia/Shutterstock.com; **181** okili77/Shutterstock.com; **183**i Gwen Shockey/Science Source; **185**, **186** Michaelpuche/Shutterstock.com; **187**t phoelixDE /Shutterstock.com; **187**b, **188** Maridav/Shutterstock.com

Chapter 16

189 shaineast/Shutterstock.com; **193**, **194** Alexander_P /Shutterstock.com; **195** Alessandro Colle/Shutterstock.com; **196** Itan1409/Shutterstock.com; **197** DenisNata/Shutterstock.com; **198** © iStock.com/Stígur Már Karlsson/Heimsmyndir

Chapter 17

201 Aspen Photo/Shutterstock.com; **203** Valentin Valkov /Shutterstock.com; **205** Rawpixel.com/Shutterstock.com; **207** Fascinadora/Shutterstock.com; **208** Africa Studio/Shutterstock .com; **209** Veres Production/Shutterstock.com; **211**t thinz /Shutterstock.com; **211**b Luis Molinero/Shutterstock.com; **212**l Ioannis Pantzi/Shutterstock.com; **212**r VaLiza/Shutterstock.com

Chapter 18

213 © iStock.com/HenkBentlage; **215**tl © iStock.com /benemaleistock; **215**ctl © iStock.com/dennisvdw; **215**cbl © iStock.com/Gary Kavanagh; **215**bl © iStock.com /SimonDannhauer; **215**tr © iStock.com/donald_gruener; **215**ctr Alexandra Giese/Shutterstock.com; **215**cbr © iStock.com /mdr0910; **215**br © iStock.com/Rocky89; **216** Sven Hansche /Shutterstock.com; **217** JTKPHOTO/Shutterstock.com; **218** StoneMonkeyswk/Shutterstock.com; **219** surasak khankasikam /Shutterstock.com

Chapter 19

223–24 starryvoyage/Shutterstock.com; **225** 1tomm/Shutterstock .com; **227** © iStock.com/TheCrimsonMonkey; **230** Rawpixel .com/Shutterstock.com; **231** Zen Shui/Odilon Dimier/Media Bakery; **232** Aleksei Kazachok/Shutterstock.com; **233** Michel Godimus/Shutterstock.com

Chapter 20

235t © iStock.com/ansonmiao; **235**cl Martin Shields/Alamy Stock Photo; **235**c © iStock.com/apomares; **235**cr Ryan McGinnis /Moment/Getty Images; **235**bl © iStock.com/swissmediavision; **235**br Boris Sosnovyy/Shutterstock.com; **236**t Gino Santa Maria /Shutterstock.com; **236**ct All Canada Photos/Alamy Stock Photo; **236**cb © iStock.com/kozmoat98; **236**b rsooll/Shutterstock.com; **237** © iStock.com/epicurean; **238** Matt Gibson/Shutterstock.com; **239** Universal Images Group/Getty Images; **240** peng_ch /Shutterstock.com; **241** dvoevnore/RealVector/Shutterstock.com; **242**, **243** Yury Zap/Shutterstock.com; **245** Gladskikh Tatiana /Shutterstock.com; **247** Sirichai Puangsuwan/Shutterstock.com

Back Matter

250 Natasa Re/Shutterstock.com; **251**t David Tadevosian /Shutterstock.com; **251**b Flashon Studio/Shutterstock.com; **254**t Middle Temple Library/Science Source; **254**b Vladislav T. Jirousek/Shutterstock.com; **258** Lebendkulturen.de/Shutterstock .com